Advance Prai

"Michael Brownlee confronts the reality of our impending food crisis like nobody else I know. He delivers the hard-nosed facts about the inevitable collapse of our global food system–sugar-coating and sparing nothing–and then in soaring, absolutely poetic language calls us to the critical work of rebuilding our local foodsheds, yard by yard, farm by farm, locality by locality. If you like to eat, and want to continue doing so, read this book by America's leading voice on food localization. It will wake you up and stir you to start working while there's still time to prepare."

—Tim Rinne, founder, Hawley Hamlet, Lincoln NE

"In our culture, local food is still largely regarded as a kind of hippie phenomenon which will always be only a tiny part of our 'real' industrial food system. But in Michael Brownlee's new book, we are introduced to an altogether new description of the future of food, and how some of the many challenges that will evolve in the decades ahead will necessarily transform our food system into a kind of bioregionalism that is grounded in our local social and ecological communities. This local food revolution promises to provide us with a healthier, more resilient, regenerative food system that will be essential to our ability to feed ourselves in our not-too-distant challenging future. Interestingly, author John Thackara confirms that such bioregional economies, including food systems, are in fact already beginning to emerge throughout the world (*How to Thrive in the Next Economy*, 2015). So Brownlee's book is one that anyone interested in the future of food should read!"

—Frederick Kirschenmann, President, Stone Barns Center for Food & Agriculture; Distinguished Fellow, Leopold Center for Sustainable Agriculture; author of *Cultivating an Ecological Conscience: Essays From a Farmer Philosopher*

"Read this book. Regenerating local food networks is one of the most crucial tasks of our time if we are to restore human health, climate stability, and the environment."

—Ronnie Cummins, Director, Organic Consumers Association

"Once again Michael Brownlee has crafted a meticulous guidebook to heal our disconnected food system. In *Reclaiming the Future*, his compassionate yet brutally honest voice taps into an ancient consciousness that awakens an emerging fire within, demanding immediate change. This voice is our moral compass. In reclaiming our regional foodsheds we are not only planting seeds of hope, but we are establishing the salvation and freedom of our global tribe. This is a breathtaking work of tremendous magnitude and impact that will resonate for generations into the future."

—Chef Daniel Asher, River and Woods Restaurant

"Michael Brownlee thinks deeply about where our food should be coming from, and has explored these dimensions for years. He opens for us a new discipline: *deep food ecology*. Read and be amazed, then determine to chart an unblazed path as Michael has done."

—Gary Paul Nabhan, Franciscan Brother; W.K. Kellogg Chair in Southwest Borderlands Food and Water Security, University of Arizona; author of *Growing Food in a Hotter, Drier Land: Lessons from Desert Farmers on Adapting to Climate Uncertainty* (and more than 30 other books)

"If you feel spiritually called to be involved in the local food movement, but aren't sure about your place within it, you need to read Michael Brownlee's book. *Reclaiming the Future* is based on the premise that local foods are not a food fad or niche market but reflect the emergence of inevitable evolutionary change that ultimately will transform the entire food system. He writes, 'the first three stages of the evolution of modern civilization have brought us to the brink of global disaster.' The local food revolution is a logical response to this crisis. It's not easy to be deliberate in the midst of a crisis, but that is exactly what Brownlee asks us to do. He quotes Martin Luther King, Jr., 'The arc of the moral universe is long, but it bends toward justice.' Brownlee believes many are called to be 'catalysts' of the local food revolution, but must be patient. We can be facilitators of the great food transformation but cannot make it happen. We simply need to fulfill our purpose as responsible members of the moral universe as it evolves toward a new era of individual and community betterment. Brownlee believes many of us will find our purpose within the local food movement."

—**John Ikerd**, Professor Emeritus, Agricultural and Applied Economics, University of Missouri; author of *Revolution of the Middle...and the Pursuit of Happiness* and *A Return to Common Sense*

"Restoring our relationships to food, the land, and each other is long-term difficult work. Michael Brownlee has demonstrated the determination and creativity needed to improve our local food systems. His new book, *Reclaiming the Future*, delivers important lessons learned and approaches that can be used by communities across the U.S. to take food system relocalization into their own hands."

—**Dan Hobbs**, Arkansas Valley Organic Growers, Rocky Mountain Farmers Union

"Brownlee boldly confronts the tsunami of damage to human and planetary well-being stemming from the dominant food system and its associated ideologies. Following Teilhard de Chardin and Thomas Berry, the author anchors all hope on our individual agency as co-creators in the universal evolutionary project. It is a project of blind faith in a race against business as usual. With everything to lose, a regenerative revolution begins."

—Bruce Milne, Director, Sustainability Studies Program,
University of New Mexico

"Beyond imparting facts, beyond organizing a movement, beyond an engraved invitation to activism, *Reclaiming the Future* is a clarion call not only to forge radical local food movements, but to a spiritual awakening that has literally erupted from the Earth in the sacred act of eating. On the one hand, freedom and human sovereignty are at stake. On the other hand, through involvement with conscious, vibrant local food movements, we are being offered the opportunity to immerse ourselves in our 'spiritual DNA' in order to rediscover who we really are and what we really came here to do. Reclaim your food, reclaim your future."

—Carolyn Baker, author of *Savage Grace: Living Resiliently in the Dark Night of the Globe* (with Andrew Harvey) and *Collapsing Consciously: Transformative Truths for Turbulent Times.*

"The time has come to commit to the local food movement–to give folks the tools to grow their own food in their own communities, and to take control of what has become an unsustainable way of producing food. The steps are simple, and the longing to return to something more tangible–that you can feel, smell, and touch–will make the local food movement thrive."

—Carrie Balkcom, founder, American Grassfed Association

"In *Reclaiming the Future*, Michael Brownlee issues a distinctive call to all those who have ears to hear to become local food revolutionaries and evolutionary catalysts. Ten years ago, at a critical moment in my own development, he did the same for me, and I have never looked back. If you pick up this book, be prepared for the adventure of a lifetime!"

—Don Hall, Co-Director, Transition US

"In *Reclaiming the Future*, Brownlee sends out a revolutionary call to action for citizens to unite and reclaim our food security, food sovereignty, and our connections to life and one another. Reconnecting to our food by growing healthy and regenerative local and regional foodsheds that promote life rather than destroy life is imperative. For those who sense this call to action but don't know where to start, this book is a must read."

—Tom Abood, founder, Local Matters Investments

"There is not a single nail that Brownlee does not squarely hit on the head in this new book. It is painfully on-point and relevant for me given where I am spiritually, professionally, and in terms of the commitment I am making to my community. Specifically, I do not wish to be an activist, but the description of a *local food catalyst* resonates perfectly."

—Jeffrey Potent, Adjunct Professor, International and Public Affairs, The Earth Institute, Columbia University

"*Reclaiming the Future* is bold guidebook for personal and cultural transformation that begins with food. Why food? Because the industrial food system is starving us of connection with ourselves, each other and the earth. Eating local food is a revolutionary act because it restores our place in the world and is a gateway to living in strong community, where the bonds of life are understood and celebrated.

Reclaiming the Future is no less than a guide to self-realization into a life of beauty and belonging. We have a long way to go—this is the work of generations. So, how do we begin? This revolution starts with you, with every bite of food you eat, every farmer you support, every market you engage. *Vive la revolution regenerative!*"

—**Philip Taylor**, Environmental Studies, University of Colorado; founder of Mad Agriculture

Taking Back Our Food Supply

TAKING BACK OUR
FOOD
SUPPLY

How to Lead the
LOCAL FOOD REVOLUTION
to Reclaim a Healthy Future

MICHAEL BROWNLEE

NEW YORK

LONDON • NASHVILLE • MELBOURNE • VANCOUVER

Taking Back Our Food Supply

How to Lead the Local Food Revolution to Reclaim a Healthy Future

© 2019 Michael Brownlee

Published in New York, New York, by Morgan James Publishing. Morgan James is a trademark of Morgan James, LLC. www.MorganJamesPublishing.com

ISBN 9781642791624 paperback
ISBN 9781642791631 eBook
Library of Congress Control Number: 2018907636

Cover Design by:
Rachel Lopez
www.r2cdesign.com

Interior Design by:
Chris Treccani
www.3dogcreative.net

Cover Photo by:
Cynthia Torres

Morgan James is a proud partner of Habitat for Humanity Peninsula and Greater Williamsburg. Partners in building since 2006.

Get involved today! Visit
MorganJamesPublishing.com/giving-back

To the Messenger and the ancient lineage he represents,
the very angels of evolution.

Table of Contents

Chapter 1

Introduction

Annalee Gibbons was proud to be part of the local food movement, and glad to be invited to a day-long regional gathering of farmers, local food entrepreneurs, policy makers, and investors.

When she arrived at the Grange Hall a few miles outside of town, on a beautiful fall day, Annalee was struck by how old the building was, and how it seemed to be a relic from the people's movement of the 1890s, organized around an alliance of farmers who rose up against corporate capitalist control of both the Republican and Democratic parties and the co-opting of the land economy.

Her own family had been farming in those days, and while she knew only a little of their situation, she felt a deep camaraderie with her forebears who were undoubtedly part of that early grassroots uprising. She was now a fifth-generation farmer on the family's land, struggling to recover and preserve a tradition and a way of life that had nearly been lost and was rapidly vanishing around the country.

Her daughter Summer had reluctantly agreed to come with her to this gathering, and as they joined the ranks filing into the meeting hall she said, "Folks here don't look very happy, Mom."

Summer was newly pregnant with her first child, and Annalee at age 59 was soon to become a grandmother. This was both exciting and worrisome, because it meant that for a while the family would have to cut back on their farming operations.

Annalee's daughter had taken the lead role on expanding the family farm, and was her best hope that the farm would have a future. But now that future seemed cloudy.

—

When Annalee's husband Samuel had contracted cancer seven years earlier, it came as quite a shock. It was a good thing her career as a partner in an accounting firm in the city had been so solid, because without the income from Samuel's construction job, they would have been in a real bind. He was gone in less than two years from the time he was diagnosed.

They had been living on the family farm for all of their married life, but had given up dairy farming many years ago when they got squeezed out of the market by the big commercial dairies. It was hard for them to let go of farming, but at least they had managed to hang onto the property.

Annalee had always been very passionate about farming, especially raising grassfed animals–heritage breeds of cows, sheep, pigs, and chickens. She was a good gardener, taught by her mother, and they always had wonderful vegetables at home.

When at about ten years old, Summer suddenly started developing serious allergies and asthma, Annalee started reading everything she could find to discover why this might be happening. She learned that the chemicals and hormones that were used in the form of agriculture they and their neighbors had adopted were severely damaging to human health. She was horrified, and made the decision to raise her daughter on only organic food.

Along the way, she got very interested in other farmers who were struggling to make this transition. They all wanted to feed their families healthy organic food, raised on their own farms as much as possible, and they were running into numerous challenges to get there.

Not long before her husband died, Annalee had persuaded him to join her in a small group of city people learning how to invest together in local food and farming enterprises, providing patient "nurture capital." It seemed like a great way to learn about the future of family farming, and to understand the challenges others were having in the burgeoning local and organic food movement.

Annalee became very passionate about supporting local food and farming. She was especially interested in the connection between nutrition and health, and how to produce the healthiest food that would be the best for her family.

She hadn't thought much about it before, but after Samuel died, Annalee found that she was suddenly free to embark on a new phase of life. She decided that it was time to make a bigger commitment to farming *organically*. She bought 140 acres adjacent to the old family farm—on the strength of her daughter's growing interest in becoming a farmer herself—expanding their ability to raise grassfed livestock.

Summer particularly loved the animals, and fortunately had partnered with a young man who was willing to learn to farm. Annalee set them up on the new acreage, which brought great joy to her life and helped ease the pain of Samuel's untimely passing. They would be able to produce most of their own feed for the animals, and thereby keep the entire operation organic.

Annalee's financially rewarding career as a partner in the accounting firm had made it possible for her to take the risk of buying farmland. The land was relatively affordable, because it was out in the country, far enough from the city that the developers hadn't yet driven up the price per acre.

The local food movement in the region seemed to be growing rapidly, and Annalee was hopeful that her daughter and her family might be able to carve out a very profitable niche in this emerging market.

No one in her nurture capital group realized just how important farming was to Annalee personally and to her family. They were surprised to hear she had invested in farmland, and even more surprised when she bought a second 140-acre parcel about a year later.

But Annalee felt she was seeing the future and responding to it. She was thrilled to be putting her money and her energies into this effort, and confident that Summer and her husband were on their way to a real career. It wouldn't be easy, of course, but Annalee felt they were forging a pathway to a future that also preserved family farming in an increasingly urbanized landscape.

The family's property was located near a small rural community, about an hour's drive south of the city where Annalee's office is located–that is, when the weather was good. They were a few miles off the highway, and sometimes the unpaved road didn't allow Annalee to get in and out, even in her rugged SUV. Too much snow or too much mud could leave them isolated for days at a time.

Living on the farm and having her main business in the city gave Annalee a real sense of the rural/urban divide that seemed to be tearing the nation apart. Sometimes she felt that she was leading a schizophrenic life, split between the land and the city.

As she learned more and more about what was going on around local food and farming, Annalee got more personally involved with some of the businesses that were emerging around this movement.

She experimented with investing directly in a small number of entrepreneurial operations, and served as a business advisor and CPA for others. Some of those investments have done well, and others have not.

But as their family's farm operation grew, and as Summer and her husband were able to devote their full-time effort to farming, they started running into problems. They couldn't find enough people who were willing to consistently buy what they produced on their farm–primarily grassfed meats, some of the best around–*and* who were willing to pay them a fair price. They were too small to satisfy the big commercial buyers and institutions, and too isolated to make a meat CSA work.

Annalee began to wonder if there really was a market for the kind of farming they were doing. And when she investigated what food buyers and restaurateurs were actually purchasing, and what was available at farmers markets, she discovered that much of what was advertised as local and organic was actually not. This was shocking to her, and depressing.

The more research she did on this, the more she learned that the local food system is full of misrepresentations and questionable practices. This is very damaging when you're trying to present an image of grassroots authenticity.

In her investment work, Annalee had become increasingly outspoken, advocating for local food and farming. But as she began to understand how the local food movement was actually shaping up as an industry in the region, she had to admit that she was becoming a little disheartened.

Some of the food-related businesses Annalee invested in were having a hard time making ends meet. The prices they needed to charge to become profitable were perceived as too high (especially compared to cheap industrial food). She was seeing some of her favorite people–including clients–go out of business. Things were not all roses in the local food movement, at least not this region, and she just didn't see how anything could change significantly, at least not in time to be ahead of the coming impacts of global warming.

All this left Annalee in a quandary, not really knowing where to turn. The situation was particularly painful for her because her understanding of the importance of local food and farming had increased significantly, and she was coming to see that food localization was an absolutely necessary approach to the challenges our world faces with global warming, economic disparity, political conflict, and local resilience.

Annalee was committed to being a farmer, to being a nurture-capital investor, and to helping others build successful businesses in this movement. But she was worried that the movement had a much tougher row to hoe ahead than she had ever realized before.

The family's plan was for Summer and her husband to build a life on the family farm, and thereby make it possible for the farm to return to its true heritage. But as Annalee and Summer walked into the Grange Hall, that heritage was now in doubt. They were struggling to find a steady local market for the grassfed meat their farm produced.

"I have to find a real market for local food," Annalee thought, "or I just don't know what will happen to our farm and Summer's future."

—

Summer and Annalee watched as people pulled their folding chairs into a circle in the auditorium, the bare wooden floor creaking with age. They sat down on the outer perimeter of the circle, where they could see and hear everyone.

The gathering had been organized by Jim Ruskin, a young farmer who was trying to figure out how to increase production on his small organic vegetable farm. The soil was somewhat depleted, and he wasn't having a lot of success.

The purpose of the day's meeting was to explore the state of food and farming in the area, and to see how local investors might be able to help by making capital available.

But as she listened to the discussion, Annalee felt her concerns grow.

Farmers reported that sales at the farmers markets were beginning to decline. This was bad news for the 80 percent of market farmers who had to work off-farm jobs in order to be able to afford to continue farming.

Growth of community supported agriculture (CSA) over the last few years had been encouraging, but now was flattening out.

While commercial food buyers complained that there was insufficient supply of local food to meet growing customer demand, many farmers were finding it difficult to sell their products profitably. Their best margins came from selling directly to people through farm stands, CSAs, and farmers markets. But those avenues didn't show much promise for growth in the future.

Farmers seemed to agree that there was a lot of resistance to the prices they had to charge to try to make ends meet. They said that people often complained that the price of local organic food was way too high, unaffordable for all but the financially blessed.

An egg producer told the story of being frustrated that he could never quite hit profitability. His organic pasture-raised eggs sold fairly well at $6 per dozen, but no matter how much he sold he just couldn't make any money. He told the group, "I had a background in financial management, and I'm supposed to be smart with numbers, so I decided to do a very rigorous cost analysis. What I discovered was that it was costing me $6 a dozen to produce those eggs. No wonder I wasn't making it."

Annalee and Summer glanced at each other. None of this was a surprise. They knew how expensive it was to feed and raise chickens organically. They also raised heritage turkeys, and were having trouble finding customers willing to pay the prices they had to charge to make the effort worthwhile.

—

Many of the things she and Summer heard at the Grange that day were disturbing.

"It's getting harder to do this work, not easier."

"Farming is an overwhelming challenge."

"Farms around here are not profitable. I live on next to nothing."

"The CSA market may have peaked. Now there are too many markets."

"Wholesale competition is tough, and it's hard to make ends meet at wholesale prices, especially when buyers are trying to drive prices down, treating local food like a commodity."

"Access to land is becoming a real problem. Who can afford land these days?"

"Finding farm labor is a huge challenge, a real limit to our ability to grow."

"Actually getting food to people, finding the ways to reach them, is the key problem. We don't have the answers yet."

"We're not replacing our older farmers. Those farmers over 65 years old now outnumber farmers under 35 by a ratio of six-to-one."

"Some of our farmers are abusing their CSA members with 'seconds,' saving their best produce for restaurants or the farmers market where appearance and quality are in demand. It's not right."

"It's hard to survive without any kind of crop insurance."

"County land use policies are preventing us from growing our farm businesses."

—

By mid-afternoon, Summer had heard enough. "Mom, let's get out of here. These people are clueless."

Annalee was a little surprised by her daughter's brusque tone, but had also reached the point that she wanted to leave.

They didn't say goodbye to anyone, but quietly slipped out the back of the Grange Hall.

As they started their long drive back to the farm, Summer curled up in the back seat of the SUV to take a nap. The silence gave Annalee an opportunity to think through what they had experienced at the Grange.

Maybe we've hit the wall. The movement seems to be running out of energy. There's such an undercurrent of frustration and disappointment. It's a sad state of affairs. This is turning out to be much harder than anyone ever expected.

She was very familiar with all the issues that were being discussed, and she had a lot of compassion for those who were there and were willing to speak up about what was going on.

Well, we clearly haven't gotten as far in the local food movement as we had hoped we would by now.

It was not that what people were saying was wrong. They were merely expressing the obstacles they were running into. They just couldn't see how all the pieces needed to fit together—access to land, access to labor, access to local capital, policies that support local food, and infrastructure that can tie everything together into a system.

Then it dawned on her. There had been something fundamental missing in the conversation at the Grange. No one was talking about *why* local food and farming is important. No one had mentioned the need to localize the food supply, that this was a matter of food sovereignty and food security.

Instead, everyone was concerned about their own operation, their own business, their own market, their own problems. But they were all operating in silos. They just didn't seem to realize that they were all working on one single project, building a viable local food system for their region.

Something needs to happen, or we're going to see more business failures. We're going to lose more farmers. But I don't really know exactly what should happen, or what my role in all this should be.

And if I can't find a market for what we really want to produce on our family farm, then my daughter may not have the kind of career she really wants. I don't know if our farm can survive for very long under these circumstances. It just may not be financially sustainable.

Annalee was feeling something stirring in the pit of her stomach that she was afraid she was not going to be able to ignore for much longer.

But surely there is a way for all this to work, so that people can have access to locally grown, healthy food, and that farmers can do what they do best and love most–feeding people, friends and neighbors, and supplying local businesses.

What should I do? I've learned so much over the last few years. I'd love to be able to leave my CPA position in the firm (but maybe hang on to a few favorite clients) and devote more of my time and energies to building our farm and building the local food movement. But now I'm beginning to fear that might never be possible.

—

What Annalee didn't yet realize was that she was about to be plunged into the adventure of a lifetime, and that she was going to have to make some very tough decisions.

A few days later, Annalee found herself in a city bookstore holding something she had never noticed before, a small book calling for a local food revolution. As she opened it, she was thunderstruck to read:

When we grasp the realities of our predicament, then we simply have no alternative but to quickly build our own local food supply.

It must start right in our own communities, right where we live. This must become the highest and most urgent priority of our communities. Our children's lives depend on us doing this!

But who will do this? Who will mobilize this revolution?

And here is perhaps the worst news of all. There is no one to mobilize this revolution in your community but *you*. You can't look to anyone else to do this. *This falls to you.*

If you can begin to see this and feel this, then you will realize that you have no choice but to lead your people to rise up and take back their food supply.

The transition to a localized food and farming system–a regional foodshed–means that nearly every aspect of the way we feed our local populations must be redesigned and rebuilt, from the ground up. And it all begins with you.

Her hands trembled as she opened up the next chapter.

A real, regenerative revolution is a rare thing. It asks something of us, and ultimately it will not be possible for you or any of us to remain neutral.

Regenerative revolutions are not based on economics, politics, or ideologies. Fundamentally, they are driven by moral, ethical, and even spiritual values. A regenerative revolution is called into being. And it calls you to a deeper level of being, and you cannot escape.

The local food revolution is a calling for you to rise to the greatest occasion in human history, a calling that will radically reorganize your intentions, your priorities, and your actions.

The local food revolution is the pathway. Highly localized regional foodsheds will be our lifeboats through the stormy seas ahead.

Now it's your turn. You are greatly needed.

—

Annalee's legs were shaking by the time she got to the cash register to pay for this little book. Her heart was racing as she finally drove onto the dirt road leading to the family farm. She wouldn't say anything to her daughter just yet, but Annalee knew that she was going to have to spend some serious time with this message.

She was grateful that the season's harvest was over and there was plenty of feed in the barn, giving her time to read and think.

This was going to be a long and life-changing winter.

——————— **Chapter 2** ———————

Why This Book?

This book is intended for people who, like Annalee Gibbons, understand they need to help accelerate and expand the local food movement in their community but don't know how–in other words, the people who are *called* to this work.

For instance, you might be (or becoming)

- a farmer or rancher needing to grow your operation or reach new markets
- a local food entrepreneur seeking customers or capital
- a local food activist or organizer striving for greater impact and community engagement
- a policy-maker looking for innovative ways to support local food
- a chef or restaurateur looking for creative avenues to increase customer loyalty through local sourcing
- a commercial food buyer trying to figure out how to meet customer demand for local food

- or even a "foodie" who is drawn to move beyond being a mere consumer of local food and wants to get more involved in the movement

The point is that whatever your current relationship to local food or agriculture, somehow you know you need to help radically accelerate and expand the local food movement in your community, *but you don't know how to do it*.

The book that you're now reading is a refined version of the one that Annalee Gibbons would eventually find, propelling her into a whole new phase of life.

Annalee's search led her to the discovery that she was destined for a role in the unfolding local food revolution far beyond anything she had imagined before.

She would indeed go on to forge pathways to new markets for her family's grassfed meat products, but she would get there through an unexpected journey, by becoming a respected and effective leader of the local food revolution in her community.

My deepest desire is that this book will do exactly the same for you, empowering you to achieve your goals in local food work, whatever they may be, by beginning to take your place as a leader of the local food revolution in *your* community.

Rethinking the local food movement

One of the great challenges of the local food movement today is that local food has primarily become seen as an expensive but optional lifestyle choice for the economically privileged (and mostly white). While local food obviously has significant benefits–to human health, to the local economy, and to building a sense of community–it's just not thought of as terribly important in our communities. It's considered a kind of luxury–something like expensive exercise equipment, or high-priced clothing. It's nice if you can afford it, but it's just not a very high priority anywhere.

After having worked in this arena for the last twelve years, I have come to understand that this is a fundamental problem in the local food movement that is holding us all back.

Perhaps I should pause a moment here to say something about who I am in all this, and who I am to you.

From the time my partner Lynette Marie Hanthorn and I began this work in Boulder in 2005–after a shattering and life-changing encounter with a spiritual teacher based here who set us on this path–we've understood that if our community was going to have any possibility of being resilient and self-reliant in the face of converging global crises, we would need to quickly learn to meet our essential needs as locally as possible.

It was soon obvious to us that food was going to be the crucial issue, our greatest area of vulnerability. And we also understood that at first almost no one else would be able to see the need to localize the food supply.

So we set to work, not fully knowing how great the challenge was going to be, or really even understanding just how important it was that our work would ultimately be successful. (You'll meet Lynette Marie later in this book.)

Over the years, we've seen local food become very popular, very trendy, and certainly very expensive (at least compared to industrial food). We've witnessed local food becoming seen as a lifestyle movement for the economically well-to-do, and people like us being accused of merely being elitist foodies.

This is one of the underlying reasons why it's been so difficult to mobilize significant support for the local food movement. While lots of people were *in favor* of local food–that is, they weren't exactly opposed to it–I began seeing that we *had failed* in communicating just how important it is to localize our food supply, even though

- we had explained the health benefits of local food
- we had explained at length the economic benefits to communities
- we had even demonstrated the renaissance of community that local food and agriculture could create

But none of this was compelling. As a result, local food has only been nibbling at the edges.

I can tell you that this has been extremely frustrating to our farmers, who are dedicated to feeding their communities with the healthiest and most nutritious food available anywhere, taking enormous financial risks year after year, and even working off-farm jobs to afford to be able to do this.

And this has been extremely frustrating to those of us working on the front lines of the local food movement. Our organization originally proclaimed a goal of 25 percent local food in Colorado over a decade. More than a decade later, we see that despite all the great local food work being done across the state we've hardly moved the needle at all. We're still at no more than 1.5 or 2 percent local food here. That's dismaying.

The truth is, our communities have simply had no compelling reason for localizing their food supply.

But that's about to change. Together, we're going to fundamentally change the conversation about local food! We're going to mobilize a revolution and we're going to ask you to play an important role.

Radicalization, transformation

Together, Lynette Marie and I have done our homework, and have been deeply involved in the front lines of the local food movement for many years now.

Along the way we've been offering a steady stream of hard-won perspectives, insights, and innovative approaches.

We've written a book all about the local food revolution, published in October 2016, which some readers suggest will be a classic for generations of those involved in this work.

Based on that book, we've developed an online course to help guide emerging revolutionary foodshed catalysts, which we released in January 2017.

And with the Local Food Summit, which we launched in August 2017, we've assembled the beginnings of a significant educational and transformational resource.

In recent years, we've said a great deal about what needs to happen in this local food revolution. But now we're stepping up to a whole new level of engagement.

—

Since I began speaking of the need for a local food revolution a few years ago, I've been going through a process of profound transformation, a revolutionary radicalization–but reluctantly. Frankly, I've been resisting. I've been denying what's been welling up inside me. But no longer.

Recently I finally reached the point where I knew that I could no longer speak and write about the need for the local food revolution without being willing to actually lead it–that is, to serve those who will be the leaders in this local food revolution. That is, *to be in the trenches with you.*

In this book, I'm going to invite you to join me in this process of radical transformation. Going through this process is a radical act of love–for yourself, and for the life that is yearning to emerge on this planet.

A revolutionary moment

Late this summer, after the extremely intense experience of producing the Local Food Summit, with more than 90 presentations and interviews with leaders in the local food movement, we decided to take a few days of retreat in Santa Fe, to allow ourselves time to rest and renew.

I had been feeling an internal pressure building for many months, but I could not quite name it. I felt there was a book wanting to come through me, and I knew it would have something to do with the local food revolution. But nothing was clear. And I sensed that more was emerging than I could currently see.

I've had enough experience with visioning to know that there's no forcing these things. I had to create the space and environment in which clarity could appear. Thus, we gave ourselves the gift of a rare retreat in Santa Fe.

While there, we noticed that Al Gore's new film, *An Inconvenient Sequel*, was playing at a theater not far from where we were staying. I had been disappointed to miss the film when it first opened, but knew I would have to wait until the Summit was over to see it. Now was the time to finally catch up with Al Gore.

Honestly, my expectations were low. Gore–who likes to quip that he was formerly the next President of the United States–had gradually faded from our radar since his original film and accompanying book, *An Inconvenient Truth* went viral in activist networks across the country in 2006.

We had been impressed at Gore's attempt to change the public conversation about global warming. He was determined to raise awareness and inspire action on a broad scale. For years, he had been developing a powerful presentation, illustrated with elaborate PowerPoint slides. He gave hundreds of talks, evolving his presentation as he went, incorporating all the latest research. But he knew he was just one voice out there, and that he needed to find a way to get his message to many more people. So he created *An Inconvenient Truth.*

But while the film and the book were very powerful, arguably the best presentation of the issues of climate change available anywhere, Gore didn't stop there. He invited activists to work directly with him to learn how to give his presentation, and to become part of a highly-trained cadre of people who could take his message out into the world, starting in their communities, and ultimately to policy makers. Since 2006, many thousands of people all over the world have been trained by Gore, and together they have had significant impact–including taking his message all the way to the Paris climate talks in 2015 (which is what *An Inconvenient Sequel* is about).

Say what you will about Al Gore, and while it's true that his message has become watered down over the years, this was a brilliant strategy.

Disrupting the conversation

What Gore did is part of a time-honored tradition of galvanizing and training grassroots speakers and activists. Richard Heinberg, senior fellow at Post Carbon Institute, told us years ago the story of the Populist movement in the 1890s, which saw the emergence of a Farmers' Alliance who rose up against corporate capitalist control. They trained an army of thousands of speakers to fan out across the country, mostly on horseback, to shift public awareness about the dangers of corporate dominance of politics and economics.

In that little underground theater in Santa Fe, for the first time it began to dawn on me that what we were doing with local food work was structurally very similar to Al Gore's strategy. This gave me a new way to think about what needed to happen now.

I knew that we needed to fundamentally shift the public conversation about local food. I knew that it was necessary to catalyze a radical acceleration and expansion of the local food movement in impact, effectiveness, and scale. But what suddenly began to open up was a vision of taking our work to a new level.

We're hoping that you are part of that vision.

The book you are reading now resulted directly from that experience in Santa Fe, and is intended as our equivalent of *An Inconvenient Truth*, the first step toward fundamentally shifting the conversation about local food, the first step toward igniting a local food revolution.

And as you might suspect by now, this book is also a recruiting tool to enlist you and others who are similarly called to become leaders of the local food revolution in their communities.

And finally, this is a book that you can use to recruit others into the local food revolution. Everything you need to do this is included.

Making the case

In this book, I'll make the case that localizing our food supply–taking back our food security and food sovereignty by building highly localized regional foodsheds–is the most important and most urgent need in the world today.

I'll make the case that the global industrial food system–the largest and most destructive industry in the world–has stolen our food supply, has stolen our health and well-being, has degraded our biosphere, has stolen our freedom, has stolen our wealth, has stolen our democracy, and is stealing our future right before our eyes.

I'll make the case that highly localized regional foodsheds can be humanity's lifeboat through the stormy seas of an uncertain future to an epoch of restoration and regeneration, and that there is no other structure or strategy that can do this.

I'll make the case that what is called for now is nothing less than a local food revolution, and that this must become the highest and most urgent priority of our communities.

I'll make the case that there are a handful of individuals who are being called to be the leaders of this local food revolution in their communities, compelled to be revolutionaries. Most importantly for you, I'll make the case that you are among them.

And I'll make the case that together we have the opportunity and responsibility to take back our food supply, to reclaim our food sovereignty and food security, to reclaim our democracy, to reclaim our freedom, and to reclaim our rapidly disappearing future.

—

This book is written for you. And it's also written as a tool that you can use to help ignite the local food revolution in the community where you live and work–a kind of book-within-a-book.

This makes it in effect a guide as to how to speak about local food, how to frame the conversation about local food in such a way that it kindles a revolutionary response.

If you apply the principles and processes in this book, you will finish with a concrete plan for preparing yourself for a mission as a local food revolutionary, a foodshed catalyst embodying the evolutionary impulse for healing, restoration, and regeneration.

This will not be easy work.

Reversing the damage

Over the last several months, it's been sinking into my core that the time we're living in now is a pivotal moment in the evolution of humanity and the evolution of the planet. It's become undeniable to me that the earth, or nature–evolution itself–is initiating a healing process on this planet and inviting us to become its agents. The process needs *us*.

The damage to life is very great, to the point that we're now facing the sixth mass extinction of species, the greatest planetary biological catastrophe since the dinosaurs were wiped out 66 million years ago–along with 70 percent of all other species. Even our own species is in danger of extinction now.

At this point, it's still possible to reverse this damage. It will take a very long time, of course, but the window of time in which we can make the crucial difference is rapidly shrinking. What we do in the next five to ten years will profoundly affect the nature of the millennia which follow. Whether humanity will even have a viable future will be determined in this very narrow window of time.

Amazingly, where the evolutionary process of healing and regeneration is beginning to emerge on this planet is around food and agriculture. It's here in the local food movement that we see the beauty of the healing process springing up out of the ground, spontaneously emerging in people's hearts and minds, spontaneously emerging from within you.

The ground has been prepared. We are being called. The spirit of healing is moving in us and through us.

It is a planetary crisis around food and agriculture that calls us–in fact, a global emergency.

Those of us who see this and feel this often feel almost completely alone. What this means is that we now bear great responsibility to respond with the very best and the very most we have to give. *This is it*.

This next five to ten years is the most crucial moment in all of human history. This is the pivot point. If we are successful, perhaps no one will remember what we are about to do. If we are not successful, perhaps there will be no one *left* to remember.

—

If you truly take it on, this book will set you on the path to the perspective, the tools, the processes, and the support that will empower you to become radically effective in the work of localizing the food supply in your community.

This is not a book about the local food revolution. It is a book designed to launch you into a life of being a leader of the local food revolution in the place where you live. *This is about you.*

If you are called to this work, this book can be a powerful way to ignite the local food revolution where it must begin–inside yourself. The revolution begins within, or it can never succeed.

You already know that you need to help radically accelerate and expand the local food movement in your community. This book is the doorway to shaping your life to become a grassroots leader in the historic local food revolution!

What's ahead

This book is designed to guide you through a transformational process, the journey of becoming a leader of the local food revolution in your community. To give you an idea of where this process is going, here I will very briefly map the seven basic stages of emerging as a foodshed catalyst–which is the structure that most of the rest of the book follows.

However, this book is not organized around information, nor is it a recipe for localizing the food supply. This is not a how-to manual. Instead, this process is meant to be primarily *experiential*, activating aspects and levels of your being *in a particular sequence* that will finally bring you to the place where you are the organic embodiment of the principles outlined in this book. I encourage you to trust the process and take your time going through the book the first time. You will come back to its gifts again and again in the future.

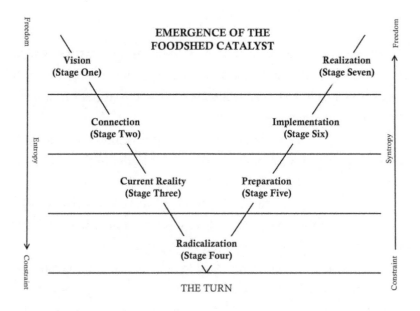

Vision (Stage One). The first stage of this process is about *seeing*. In chapter three, you will explore the vision of the highly localized regional foodshed as humanity's only viable lifeboat through a time of darkness and chaos brought about by industrial civilization. Here you will come to see yourself as part of an evolutionary course-correction. And you will begin to see the possibility of a truly regenerative revolution, beginning with food localization.

Connection (Stage Two). The second stage recounts the process of reweaving the web of connections at the heart of human community that have been torn asunder through totalitarian agriculture dominating our civilization. The theme here in chapter four is *reconnection.* You will also confront the temptation to forcefully "make things happen," and to embrace the sacredness of what is unfolding within you and around you.

Current Reality (Stage Three). In stage three, you will consider the current realities of the larger context of the work of healing and regeneration. First, in chapter five, you will have an opportunity to review the latest updates on our collective predicament as a species, and how our industrial food supply is precipitating a shattering planetary crisis. This is bad news, but necessary to take in. Then in chapter six, you will dive into the surprising *good news* of how a much-needed evolutionary course-correction is taking hold through the process of emergence and manifesting in the spontaneous uprising of food localization. Here you begin to sense where this process is ultimately going, and how you can align with and even embody it.

Radicalization (Stage Four). The turning point of the book and for you will be the fourth stage, the moment of your beginning to make the radical and unavoidable decisions and commitments that will not only shape your own future engagement as a revolutionary foodshed catalyst, but will have far-reaching impacts on millions of acres and millions of people. In chapter seven, if you choose to accept this mission, you will begin to enter the revolutionary path of the foodshed catalyst.

Preparation (Stage Five). It is in stage five that the real work begins, the long phase of preparing to gradually emerge as a leader of the local food revolution in your community. Resisting the urge to prematurely or inappropriately commit to projects and relationships that could be detours from your central purpose, in chapter eight you will begin to gain the long view of how food localization is unfolding in your region and how you might prepare to become a servant of the process.

Implementation (Stage Six). In stage six, you are finally ready to begin actually stepping into the role of leader of food localization efforts in your region. No matter what you have done before, you are embarking on something entirely new, and you will not be able to do this alone. In chapter nine, you will begin to see how you can assemble and guide a core group of foodshed catalysts with whom you will collaborate and co-create the future of food in your territory.

Realization (Stage Seven). In a sense, stage seven is about completion or fulfillment. However, the work of healing and regeneration is never really completed but keeps expanding. But at this stage, explored in chapter ten, you will gain a sense of the long road ahead of servant leadership, embodying the local food revolution in your own being. Here you will consider the lifelong legacy you are creating.

Bon voyage!

From here, I can only encourage you to read and experience this book quickly at first, ideally taking in the full scope in a single sitting, then allowing yourself time to contemplate and reflect on what is stirring within you.

You are embarking on the journey of a lifetime here, and I look forward to connecting with you in the near future as you take your place with other foodshed catalysts who are rising to the greatest occasion in human history.

#

Chapter 3

Vision (Stage One)

Prelude

We are a hungry people, malnourished and underfed.

Yet for the most part we do not suffer from a lack of food. One-third of our citizens are chronically obese. Our food is so cheap and abundant that we throw away more than a third of it–even though, as environmentalists remind us, there is no "away."

But the food that is so reliably delivered to us by the global industrial food system does not nourish us or support us. Instead, it undermines our health, degrades our environment, poisons the biota, destroys the soil, weakens our economies, and eviscerates our communities. This cannot in any meaningful sense be considered "food."

Meanwhile, we suffer from a different kind of hunger, a deeper, existential or spiritual hunger that is too often unspoken, unconscious, and ignored. Silently, we hunger for meaningful connection with the earth. We hunger for connection with the cycles and processes of nature, for connection with the sacredness of life. And we hunger for connection in community with each other.

Our global industrial food system has broken these connections.

It's not that our current food system is broken, for it functions very well indeed. It is more that it has broken our connections with the most fundamental relationships of life itself. Our food system has left us starved for life.

Our food system is breaking us. Participating in it disconnects us from living systems and from each other, unintentionally producing broken economies, broken communities, and broken people.

We could think of industrialized civilization today as a huge concentrated animal feeding operation–a CAFO–not much nicer than the massive feedlots that we subject our cattle to, or the factory farms that house and process our pigs and chickens and lambs and fish. And, for the most part, we humans are not much better off than those poor suffering creatures. We're not raised on the open range, with free access to the foods that are natural to us where we live. Instead, we're pumped full of hormones and antibiotics and fed a synthetic diet of highly processed food-like items laced with toxic chemicals and drugs designed to fatten us up for the market of the industrial health care system. We are not encouraged to be cage-free or grassfed or pasture-raised. Almost everything about our diet and our lives is artificial, controlled by transnational corporations. *We are being ranched.*

There is much anguish about food deserts these days, and rightfully so. But the uncomfortable reality is that almost all of us live in something of a food desert, where healthy, fresh, local, sustainably produced food is available to almost no one except the relatively wealthy and those who have learned to grow it for themselves.

Amazingly, most of us have been almost completely unconscious of what is happening to our food supply and what it is doing to us. For the most part, we have passively accepted our plight. However, this is quickly beginning to change.

The way we compensate for our unmet hunger is what has come to be known as *consumerism*–a social and economic order (and ideology) that encourages the ever-increasing acquisition and consumption of products and services. We have been trained even to identify ourselves as consumers–a lifestyle that emits toxic wastes that foul our air, land, and water (to say nothing of the mental and emotional toxicities that we produce along the way). While it's amazing that we are able to "live" this way at all, it is a life that does not remotely resemble the life of which we are capable–or the purpose and meaning for which we seem to be designed.

While this sad state of affairs seems to dominate our society, something new is stirring across the land–a much-needed revolution in the way humanity feeds itself. We are beginning to take back our food supply, reclaiming our food sovereignty, and building local foodsheds.

It's necessary to understand how we got into our predicament before we can begin to see the path beyond it. Here's the backstory.

Since the advent of industrial civilization, but especially since World War II, we have become increasingly dependent on a globalized, industrialized, corporatized food system, controlled by an unholy alliance of transnational corporations–known by the familiar monikers of big ag, big food, and big pharma–all empowered by big banking and big government, and fueled (not coincidentally) by big oil. This industrial food supply chain is tightly intertwined and highly focused on producing profits and consolidating power above all else.

In this system, our farmers and ranchers are told that their job is to "feed the world." But this is pure propaganda, an insidious reframing of reality–for in less than a century, this unholy alliance has successfully colonized most of the industrialized world, almost completely displacing humanity's ability to feed itself, creating a ruthless dependency for the most fundamental requirements for life.

Altogether, the industrial food system—now by far the largest and most destructive industry in the world—burns about 23 percent of global oil and gas supplies, and is apparently responsible for more than half of global greenhouse-gas emissions, and 80 percent of fresh-water resource use.

With the goal of producing commoditized food as cheaply as possible while maximizing corporate profit and control, the architects of this behemoth have put the well-being of life on this planet at risk, for it has been steadily fouling our natural environment, destroying topsoil (half of which has been lost in just the last 150 years), depleting ancient fossil-fuel and fresh-water reserves, dramatically warming our atmosphere, and gradually destroying the very conditions conducive to life. Our industrial food system has become the equivalent of mountaintop-removal coal mining—to the point that it is now an overwhelming contributor to the sixth mass extinction of species in planetary history, producing what Paul Ehrlich and his fellow researchers are now calling "a global spasm of biodiversity loss," the worst planetary crisis in sixty-six million years.

The unholy alliance has unleashed a wave of rapacious and disastrous economic growth, uprooting small farmers and converting farmlands into urban and peri-urban developments, transforming cities into human CAFOs, and making inevitable an explosion of the human population—an estimated 70 to 90 percent of whom will be forced to live in overcrowded urban environments by midcentury.

All the while, this system delivers to us food of ever-poorer quality and diminishing nutritional value, weakening our bodies and minds, causing an epidemic of diet-related diseases, decreasing our life expectancy, undermining our local economies, creating a profound economic disparity among our people, and severing our relationship with the land and with all that is sacred in life.

If left to continue, the trajectory of this juggernaut in the coming decades portends an unavoidable collapse of ecosystems, economies, and human populations.

We must recognize that the global industrial food system cannot be redeemed, for like the fossil-fuel industry on which it depends, it is so deeply corrupt and corrupting, so morally and spiritually bankrupt, that its inevitable fate is entropic decline or even widespread collapse. As Fred Bahnson says, "You can't trust Babylon with the food supply."

It is an obvious if largely unspoken reality that this food system on which we depend is profoundly unsustainable. Less obvious is the reality that it is also becoming dangerously unstable. The entire system is teetering on the brink of collapse just at the moment when it purports to be ramping up to feed a world of some nine or ten billion people by midcentury. But in fact this global food system has itself become the greatest threat to humanity's being able to feed itself, and it is already failing us. How we answer this challenge could well determine the future of humanity and even the future of life on this planet.

Agrarian roots

We are learning, thankfully, that we are by nature an agrarian people. That is, we all share deep connections with the land which provides all our food, our water, the air we breathe, our very life. But we have forgotten.

We all have agrarian or agricultural roots, and only recently have we become disconnected from them, and this has left a gaping wound in our souls. We need to heal this wound, close this gap, and reconnect.

Most of us only have to look back a generation or two–three at the most–to locate ancestors who were directly involved in agriculture. If we allow ourselves, we can still feel the connection that we have lost.

Our broken relationship with agrarian life has disrupted our connection with life itself.

All of human history is based on our connection with the land. Or disconnection. We are just beginning to understand that every previous civilization collapsed because of an agricultural crisis.

The crisis of our current industrial civilization–the first human civilization to dominate the entire planet–is that it has almost completely disconnected us from our land base, from our agrarian roots. The collapse of this civilization is now inevitable, and in fact is well under way. But what we see emerging is the very beginnings of a truly agrarian culture that will be our lifeboat to a new civilization. This is the heart of the local food revolution, cultivating the emergence of highly-localized regional foodsheds as the bridge to what "geologian" Thomas Berry calls the Ecozoic Epoch, where humanity will live in co-creative harmony with nature.

This book provides an introduction to the process of localizing our food supply, and is meant to support those individuals who feel that they are called to this work at this pivot point of human evolution, this planetary agricultural crisis, this global food crisis.

We begin with a vision of where this evolutionary course-correction is attempting to take us. We begin by seeing and feeling where we are going as a people, living in co-creative harmony with a living planet.

———

When we consider our future, we need to let go of our ideas about where we are now or how we might get to the future described here.

What we're looking for are clues to our underlying design, which will find its natural expression over time.

As a people, we're in the very early stages of a very long process of course-correction, now beginning during the darkest hour of human history.

In this world, we are witnessing the signs of collapse of human civilization. It is inevitable that this civilization should collapse, for it was built on practices fundamentally out of harmony with nature and with our underlying design.

As we will discover, these practices had to do with a particular form of agriculture that is extractive and exploitive. Daniel Quinn, author of *Ishmael*, has appropriately named this *totalitarian agriculture*, which displaces all living beings on a tract of land for the purpose of cultivating food crops (including animals) to feed humans living nearby in settlements, towns, and cities. This form of agriculture is the foundation of today's global economy, and has given rise to the vast imbalance between humans and the land.

Never mind, for the moment, how we can unravel this form of agriculture and the civilization it has spawned. Actually, as we'll see, they'll unravel on their own. Their demise is inevitable, for they are profoundly unsustainable.

—

In *Call of the Reed Warbler*, Australian farmer and ecologist Charles Massy writes of learning what life was like for the Aboriginal people before the arrival of Europeans in 1789 (nearly three hundred years later than their arrival in North America). He describes ancient place-based tribes, around 250 of them, each with their own language and culture, each with their own regionally adapted ways of feeding themselves, all of them deeply interconnected spiritually.

I know little of the history of indigenous people in North America, but I suspect that a little research will quickly reveal that the pattern of pre-Columbian life here was very similar to what Massy describes. The pattern that emerges is of tribes of people living harmoniously and sustainably within the natural limits of a bioregion, with a stable population size, caring for the land and its flora and fauna in such a way that they are easily able to feed themselves from the natural abundance of the ecosystem.

Intuitively, we know that this is how life was designed to be on a living planet. But we resist this vision, because it seems so distant. We assume this represents some remote and irrelevant past, to which we will never return.

But what I am seeing is that this way of life closely matches our description of a highly localized regional foodshed, where people are able to meet their essential needs as locally as possible. This has been the vision underlying all localization work from its beginnings some decades ago. This is not merely a nostalgic back-to-the-land movement, however. This reflects a yearning for our deepest roots, our deepest nature.

The Europeans carried with them the seeds of totalitarian agriculture, with disastrous results, bringing to an end the ancient ways of life of indigenous peoples in the Americas. This is a pattern that has been played out throughout the world.

This is the direction we're currently heading on this planet.

Based on totalitarian agriculture, this civilization took a wrong turn some 10,000 years ago, leaving its indigenous roots behind. Now, evolution is initiating the inevitable course-correction.

The mind rebels in considering what might need to happen in order to return to an agrarian way of life on this planet. We assume this would be utterly impossible, perhaps even undesirable. We assume it is naive idealism to even allow ourselves to think about such possibilities. Our disconnection from nature, from the land, from the living community of earth's biosphere, from each other, from life itself has left us bereft of the ability to envision any other form of life than what has arisen over the last 10,000 years.

We are told that the migration from the countryside to cities is inevitable, even necessary, and that as many as 80 percent of nine to ten billion people will be living in sprawling metropolises by mid-century. But this is not the future that will unfold. Instead, in the coming decades we will see a dramatic reduction of human population and a reversal of this migration. Our cities will soon begin to empty out, and people will begin to re-establish a rural agrarian life.

I realize that this is a radical vision of the future. For most of us, it's unimaginable. I'll refrain from attempting to make any specific predictions, for no one has lived through such a transition. Living at the moment of the beginnings of the collapse of our civilization, heading into a time of upheaval and chaos, we cannot imagine what life in the Ecozoic Epoch will look like, or the wrenching process we will go through in the difficult centuries and millennia of the Anthropocene. Nor do we need to engage in such imaginings.

We can learn to *feel* where we are going as a people, what our destiny is. We need not understand this, or explain it. But it is very useful to feel it.

An exercise

Take a few minutes to recall and feel your own agricultural or agrarian roots. Write down the answers to these questions.

- What are your agrarian or agricultural roots?
- How did you become disconnected from this heritage?
- What has happened to how you eat in the process?
- How could you begin to reconnect with your agrarian/agricultural roots?
- How would you want this relationship to look in the future? For your children and grandchildren?

- What would be the cost of not being able to experience this, to actually live this way?

Contemplate the following, give a few minutes to each:

- What does it mean that virtually every human civilization has collapsed because of an agricultural crisis?
- What would need to happen in order for human society to become agrarian?

A vision of ourselves

If you are reading this book, you likely have been feeling an inexplicable pull to food or farming, a sense of rightness about agrarian life, a yearning for connection to the land, a sense that localizing the food supply is a matter of purpose and integrity for you.

These are very important feelings, emerging from deep within, and they need to be cultivated and nurtured, respected, even cherished. Such feelings are your North Star as you shape your life in this time of radical change and chaos, for they are signs of the process of healing and regeneration emerging within your own being. This evolutionary process seeks to manifest in you and through you, and it is calling to you now.

You are as a seed, planted by mysterious means, destined to spring to life at this pivotal moment in human history. This is your season to break through the soil and grow to full flower. This is your season to bear fruit, to give life.

A seed does not consciously know what it will become, but the memory of its heritage is stored in its DNA. Given the right conditions–sufficient sunlight, soil, and water–the DNA will guide the plant to its full potential.

So it is for you. If you cultivate the appropriate conditions for your own emergence–sufficient love, support, and guidance–you will become the fulfillment of your spiritual DNA. This will take some time, some intention, and some effort on your part (mostly a matter of surrendering to the process). But the process will unfold as surely as the development of a corn plant tended by indigenous caretakers.

You can trust this process that is unfolding inside you, and you can trust where it is taking you. It is most important not to resist, and not to try to make it happen. You are growing and evolving, and must allow yourself to do so, no matter if you are young or old. You are in the very early stages of fulfilling your underlying purpose. You are a seedling, a sprout. It will not do for you to try to become a full-grown plant before your time. You must go through the process.

But just as we can feel where we are going as a people, it is possible now to say a few things about where you are going as an individual. You are becoming a vehicle for healing, restoration, and regeneration, an instrument of the long evolutionary arc of the moral universe. You bear within you the dream of the universe itself. This is what guides you and motivates you.

You were born for this time, and you are greatly needed. You need not be able to explain this to yourself, for any explanation or reason would be irrelevant. Your life is a mystery–full of meaning and purpose, but guided by the most mysterious and inexplicable force in the universe. You will have ample opportunity to explore this mystery, but doing so will take far longer than you might imagine now.

You are not here for your own development and accomplishment, but for what you have to contribute. You bear within you a rare and precious gift that is yours alone to deliver on this earth, at the appropriate moment, and you will spend many years preparing to be able to fulfill this function. Your life experiences have given you clues, but you cannot yet fully comprehend what your true gift is or what your specific role will be. These things can only be discovered in the process of living and becoming.

You are here to serve in a particular way. You are here to work, to give, to contribute. And you are here to join with others who have been sent here at this moment of planetary crisis to shoulder a great burden of responsibility together. The future of life on this planet depends on our finding each other and together being a vital part of the healing of the earth and of the community of life that has been so deeply damaged.

You are discovering that the emergence of regional foodsheds is evolution's primary strategy in bringing healing and regeneration to the living earth, and that you are called to be a conscious catalyst for this process–whatever your specific role may eventually turn out to be.

There are only a few of us who have found each other in this process, but we know there are many more in whom these deep feelings are awakening. You will help us find them, and you will become an elder and mentor to those who arrive later.

You are in the earliest stages of the process of becoming a leader in a revolutionary planetary course-correction. You will not be able to sit on the sidelines of this effort. You will not be able to succumb to hopelessness. You will not be overcome by self-doubt, because you know that you are not alone, that you carry ancient DNA that can come to fruition through you. You will not be derailed by the chaos that you witness, or the darkness and suffering that engulfs the lives of so many.

You see the long view. You know in your heart what is possible in this troubled world, and you know that you have a destiny to fulfill. And you know that other individuals have been similarly planted, and that you now have the opportunity to learn and grow and work together.

Comes the revolution

A revolution is emerging now, even at this moment of gathering darkness.

True revolutions only occur at the darkest hour. True revolutionaries are only called into service when it is absolutely necessary.

This is a revolution unlike anything seen before in human history. All previous revolutions failed, for they were born of anger and desperation.

What's needed now, and what's coming, is a revolution for healing, for restoration, for regeneration. This is inevitable, and it will be unstoppable.

Never before have we seen a regenerative revolution. But that's precisely the kind of revolution that is emerging now. It is not violent or even confrontational, but it is deeply subversive.

The seeds of this revolution are already among us. Now they must be planted and nurtured. This is going to be a very long-term and sustained revolution. But we will need to hurry, for the season's window of time for planting seeds is quickly narrowing.

The outward signs of this regenerative revolution will be *a radical expansion and acceleration* of the local food movement in impact, effectiveness, and scale. The inward signs will be individuals like yourself who find themselves somehow called to this extraordinary mission.

A missed opportunity

There was a moment when such a regenerative revolution could have taken place on this planet, but as powerful as that evolutionary impulse was, it was thwarted.

That time was in the winter of 1775-1776, in colonial America. This was a time of oppression of the 2.5 million colonists by the

British crown in the form of increasing violence and increasing taxation. The public debate, and the primary concern of the Colonial Congress, was finding a path of reconciliation with England. But King George III was not interested in reconciliation. He sought only more power and more taxes.

What broke the impasse was a humble 46-page pamphlet written by Thomas Paine, "Common Sense," which made it undeniably clear that the only viable path forward for the colonies was a complete break from England to form an independent nation, and that what was required to get there was a revolution calling for leadership of the kind that demanded risking everything for the sake of the future.

The revolution that "Common Sense" ignited in the space of only six months (from publication to Declaration of Independence) did indeed give birth to an independent democratic nation. But the evolutionary impulse was not completely fulfilled, for the new nation was deeply corrupted by economic greed and disparity, by its embrace of slavery, and by the decimation of the indigenous people of the land. The results of this corruption are all too visible today, along with the inevitable co-opting of democracy itself.

That same evolutionary impulse of the American Revolution is reaching a peak again today, and is the driving force behind the local food revolution. This is the vehicle that evolution has chosen to fulfill what was attempting to emerge more than 240 years ago.

This means that those of us on the front lines of the local food revolution bear within ourselves the same spiritual DNA that inspired Thomas Paine, the same drive for freedom, justice, sovereignty, and beauty. The leaders of the American Revolution are our forebears, and their spirit is in us today.

The time in which we live

We now live in an historical moment in which the conditions of the world are very similar to those in the American colonies in the winter of 1775-1776, only on a much greater scale. But this time, the deep evolutionary impulse–the long arc of the moral universe, which has been waiting for the right conditions to appear–will not be denied.

We who are joining the ranks of the local food revolution are the true regenerative revolutionaries of our time, and we gladly offer our lives, our fortunes, and our sacred honor in service to what is attempting to emerge now.

Thankfully, this revolution is well underway. Its seeds have been planted in every community, though most remain invisible. Yet we can see the seedlings of the revolution gaining strength and stamina in many communities. These are the contagious "centers of aliveness" that we will explore later in this book.

For now, what you need to know is that you are truly called to this local food revolution, and to take your place on its front lines. What this looks like for you is yet to be discovered, and you need not be in a great rush. As we will see, what comes after acknowledgment is acceptance, and then preparation. There is much to be done before we can truly live our destiny. It's a process.

—

I pray that this chapter has given you a glimpse of the future that longs to emerge at this moment in human history, and a sense of what your purpose is on this planet at this extraordinary time.

If these words cause you discomfort, that's probably a good thing. Please explore your discomfort, get to the bottom of it. Open yourself to receive the words for which your soul has yearned. Open yourself to receive the guidance that you are tempted to resist.

The intention here is to begin to open up your capacity to see and to feel, honestly and deeply. This is a discipline that if cultivated will serve you well as you unfold as a foodshed catalyst.

#

Chapter 4

Connection (Stage Two)

We are a disconnected people–disconnected from the land, from each other, from the cycles and processes of life, from our deeper nature.

Healing and restoration begin with reconnecting.

In our food localization work over the last twelve years or so, we have noticed that when people first connect with local food and farming in the place where they live, they unwittingly initiate a life-transforming process.

Often this begins with tasting real food, grown locally by identifiable farmers with whom we can actually have a conversation, discovering an abundance of flavor and nuance completely unknown to people in our culture. This usually comes as a shock, for the industrial food we've become accustomed to is lifeless and bland.

But eating local food is not just a matter of flavor, texture, and color. Yes, it tastes wonderful. But even more importantly, we experience *life* in it.

Some of us had earlier learned to appreciate the complexities of fine wine, which can give us a sense of the land on which it was produced, the methods used to grow the grapes and distill their juices, and even the soil and weather of the region of origin in a particular season. It takes some work to train our palates to be this discerning, for industrial food has nearly destroyed our culinary senses. But once our taste buds and olfactory nerves are fully awakened, informed by a basic understanding of ecosphere and geosphere, a world of gustatory delight opens to us.

So it is with local food. We discover that our senses have been deadened by our industrialized culture and the so-called food it produces. And when our senses begin to awaken, and we experience the *terroir* of local food, we become a witness to a landscape that we did not know existed–just down the road from where we live. Here we discover that there are people who dedicate their lives to producing the most wholesome, the most nutrient dense, the most health-generating and life-giving food available anywhere on the planet–and making it available to friends, neighbors, and members of their community.

If we visit a farm or ranch where this is happening, we begin to discover a vast community of living beings that co-exist in harmony with humans, from the biota of the soil to the symphony of living creatures–plant and animal–that are all part of this web of life. We simply didn't know that such collaboration and cooperation was even possible in today's world.

We also discover a community of people who are working together to make the abundant harvest of this labor–the fruits and vegetables and meats–available to us.

And we discover that living among us there is a community of eaters, people who thrive on local food and the experience it engenders.

We discover we are being introduced to the world of the local food experience, and that we have stepped onto a path of discovery that has no end. We are on our way to discovering–or perhaps rediscovering–a way of life that is so profound and so precious that we could call it a doorway to the sacred.

This is the local food experience of which we speak in our work, knowing that most people who hear these words are often confused and even a little put off by them at first. We are pointing to a living system–a living being–that is emerging in our midst, rebuilding lost connections, rekindling true community in a time of separation and isolation. What is emerging is a *living foodshed*.

Most of us have had to look hard to find our way into this local food experience. Perhaps we begin with a visit to the farmers market, where we try some unusually expensive fresh local food. Perhaps we stumble into a fine restaurant that proudly features local food. Perhaps we are invited to a farm dinner, where we experience not only local food at its finest but also get a glimpse of the richness of farm life. But however we enter into this world, it becomes very compelling. We feel that somehow it is exactly what we need.

This experience opens the way for us to become place-based, part of the natural community in the area where we live. We begin to actually experience the land, and the cycles and processes of life that occur throughout the seasons. We begin to connect what we observe in the ebb and flow of weather with what's happening on the farms and ranches that grow our food. We gain a sense of the movement of moon and sun and stars, and feel how everything is orchestrated together, deeply interrelated.

Once we begin to open in this way, life for us is markedly different from our life before, and we can see how disconnected we have become in our city-based online life. We discover what we have been missing, which activates our deep hunger for agrarian life.

We have come to live *somewhere,* and are now part of a living community that is connected with the land and the sky and the web of life. Food is at the very center of this life. We have entered into a world that had been invisible to us. Our life is being reoriented around food and farming, and we come to see and feel that all this activity is the very heart of human community. We begin to sense what it means to be truly human.

We also become aware of water, and how this precious substance makes life possible. We begin to see where the water comes from, where it goes, how it is stored, how it is wasted, how it is central to the web of life. This is no longer an intellectual understanding, but visceral. We feel our watershed and its relationship to our foodshed.

—

When we reconnect, we experience a profound sense of aliveness. We find that we are alive as never before, that our senses are full of wonder, and that we are part of something fundamental that we scarcely knew existed.

But we also become aware that very few people in the place where we live are connected to this living web, and we see how they suffer as a result, how we ourselves have suffered from this lack of connection.

If we follow our deeper impulse, we begin to want to share this experience and this orientation with others. We wish to share this life, to engage more of our neighborhood or town or city in this hidden cornucopia of abundance. It may even begin to occur to us that this way of life may somehow hold the answers to the problems and chaos of our society.

Eventually, it will occur to you that you have something to do in all this.

You may have already replaced much of the industrial food in your diet with local food. You may have become a member of a local farm, practicing community supported agriculture and experiencing the benefits of agriculture supported community. You may even have learned to grow some of your own food, joining the growing number of passionate gardeners who have become an essential part of the web of life. You may also have learned how to can and preserve the food you grow. You find yourself increasingly cooking at home, sharing local food with family and friends. You discover local food potlucks, and the community of eaters who are joining together around the joys of the local food experience.

Along the way, you feel compelled to take stronger measures. Perhaps you choose to stop eating industrial food whenever possible. You become less willing to compromise on what you take into your body. You decide that you can no longer be a passive consumer, and resign from that ignominious and pervasive role. You question the source and the quality of the food you consider buying or eating. You question farmers, chefs, cooks, servers, and grocers, and you read food package labels. You want everything you eat to be as local as possible, as fresh as possible, as close to the web of life as possible.

But as life-changing as all this is, eventually you will reach the point where this won't be enough. You will come to see that this agrarian way of life, so rich and so natural and so right, is available to only a tiny number of people anywhere. The local food experience is beyond the reach of nearly everyone except the relatively wealthy and the truly dedicated. This is corroborated in the grim realization that less than one percent of the nation's food supply is local. (The best number we have is that local food represents 0.3 percent of the entire $5.3 trillion food and beverage industry in the U.S.)

For you, this is unacceptable, even tragic. You can see that the industrial food system, which almost completely dominates the food supply, is killing us and is destroying life on this planet. You cannot live with this realization and blithely continue enjoying your own local food experience. You find yourself drawn to become more involved somehow, to become active in the local food movement.

The truth is that the local food experience is deeply activating. We feel the need to become more engaged in the shift to local food in our community, but it's troublingly difficult to find ways to become involved beyond our personal sphere. Local food is not a coherent movement.

Slow Money

Woody Tasch, a recovering venture capitalist, came to see that he might help build local food systems everywhere by encouraging individuals to invest some of their money in local food and farming enterprises. Since he launched Slow Money in 2008, he's inspired more than $60 million in low-interest loans to some 600 small farmers and local food entrepreneurs across the country. People have formed local investment clubs to explore how to do this together in their own communities.

Slow Money is an encouraging development, and it provides a unique pathway for individuals to explore how they can contribute to building the local food movement in a more meaningful and more impactful way. But this is just a beginning.

Centers of aliveness

As you pursue the local food experience and your growing desire to become more deeply involved, you will discover that a local food revolution is underway. You will inevitably encounter "centers of aliveness," where something new is breaking through that goes beyond the plethora of useful and inspiring local food projects.

In Ft. Collins, a young couple–Nic Koontz and Katy Slota–in their eighth season of farming on a patchwork of small parcels of leased land, reached a crucial point in their farming career. They had built a successful farm operation from scratch, known as Native Hill Farm, with a large CSA membership, a strong presence at the farmers market, and a popular farm stand. Unlike most small farmers, they had been so successful that they were able to buy a small home, set up 401k retirement plans, buy a $40,000 tractor, and begin their own family. They were good managers, and talented growers, and knew that they needed to consolidate and expand their operations in order to survive in a highly competitive market.

But farmland outside the fast-growing city was becoming highly valued by developers, driving up the cost of farmland to $30,000 an acre–far out of reach for almost any farmer.

They ran out of options for leasing additional land, and were growing weary of cultivating food crops on small parcels that were not even adjacent to each other.

Nic and Katy organized a community meeting, inviting their customers and their network of supporters, to declare that the kind of farming they had been practicing so successfully was in danger of vanishing. If the community could not find a way to solve the problem of access to affordable land, Nic and Katy were soon going to have to abandon farming as a viable career.

Among the audience in that emotional gathering were members of a slow money investment club, including long-time community activist Gailmarie Kimmel, who, along with a few other visionary souls, determined that they would devote themselves to finding a solution. It's a long and powerful story, the equivalent of a local food opera, but ultimately what happened was that they formed a multi-stakeholder cooperative, Poudre Valley Community Farms, whereby members of the community could pool capital (at $2500 each) to become co-owners in an organization that buys farmland and offers long-term leases to aspiring young farmers.

They are now in the process of buying their first farm, and Nic and Katy will likely become the thirty-year leaseholders on approximately 50 acres of conventional farmland that they will convert to organic food production for members of the co-op and the community at large.

Enough people are involved in this project that it's fair to say that the community of eaters in Fort Collins are determined to localize their food supply to the greatest extent possible in the shortest possible time.

The Poudre Valley Community Farms project (PVCF) is the first cooperative farmland trust of its kind in the U.S., and is being considered as a replicable model by local food activists in communities across the country. This is truly a "center of aliveness," where the emerging process of healing and regeneration represents a contagious breakthrough in localizing the food supply. PVCF is as "organic" a development as we have ever seen, arising from within the community of eaters to address a formerly intractable structural problem in the development of local food systems.

—

Small farmers in western Kansas could not find sufficient customers in their rural communities–whose population has been in decline for decades–so began organizing together to find ways to reach the burgeoning market of the Front Range of Colorado, especially Denver. There were many thorny logistical challenges to get fresh food from the farm to the hungry cities more than 200 miles away, across the rolling plains dotted with conventional farms and ranches.

The project has taken years to unfold, but today a distribution network of several farmer-owned food hubs has emerged called Tap Root Cooperative, representing more than 120 farmers and ranchers, bringing fresh local food to the Colorado urban corridor 365 days a year. This is a breakthrough in collaboration and cooperation, and has been gently facilitated by the Rocky Mountain Farmers Union, a hundred-year-old family farm organization now undergoing a vibrant renaissance driven by young farmers–under the Quaker-influenced guidance of exemplary local food farmer Dan Hobbs.

Tap Root may be the first *food hub network* of its kind in the nation. It's still in the early stages, and there are many problems to solve. For instance, these farmers have been learning that they are not necessarily good business managers. They have found accessing necessary growth capital to be a very difficult challenge on the way to financial viability. And finding the ways to actually connect with eaters who are seeking local food may be the biggest challenge of all. But together, the farmers of Tap Root are beginning to cobble together the local food infrastructure that is so badly needed by farmers everywhere.

—

In an economically blighted Denver food desert community, two young men–returning from volunteer stints in the third world that brought them face-to-face with hardships and suffering–decided they needed to find ways to reverse the downward economic spiral on their own home turf. Eric Kornacki and Joseph Teipel landed in Westwood, the most economically oppressed neighborhood in the Denver metropolis, formed a non-profit called Re: Vision, and began assisting people in growing food together. With the help of trained (and paid) *promotoras*, in a matter of a few years, more than 400 families were producing much of their own food. Recently, with city government funding, they were able to acquire rundown commercial property that was slated for re-development, which enabled leaders of the neighborhood to begin creating the Westwood Community Food Cooperative, a community-owned food hub and grocery store. Soon Westwood will no longer be considered a food desert, where the nearest supermarket was more than a mile and a half away.

Today, members of the Westwood Community Food Cooperative understand that they are building long-term community wealth, and are taking their economic destiny—and their health—into their own hands. Everything they do will be locally owned, locally controlled.

———

As the stories of these centers of aliveness spread, the ripple effects will be far-reaching. These grassroots enterprises are symbolic of how food localization is emerging as a pathway to healing and regeneration. It is a principle of evolution that centers of aliveness must generate other centers of aliveness, or they will collapse. So, these exemplary projects represent the long arc of the evolutionary force manifesting in ways that go far beyond business plans or economic development schemes. Instead, they are spontaneously arising manifestations of the local food revolution, itself part of a much larger process of reversing the damage of the industrial-growth society and ultimately making possible the end of the Anthropocene Epoch of darkness and the arrival of the Ecozoic.

———

As you study the dynamics of the emergence of these centers of aliveness, along with others that you will discover in or near your community, you will be called to find ways to initiate and even lead the development of such catalytic efforts yourself.

You will be tempted to plunge into supporting a variety of projects, from the small to the very ambitious. But you should be quite cautious in making long-term commitments of your time and resources, for you are in the early stages of discovering what your role in your community food localization effort is destined to be.

What will finally arise within you, if you allow it, is a deep inner response to a profound *specific* need or challenge in your community. This is your calling, but it may take years for it to find you and for you to yield to it. And it may be quite different from what you expected or desired it to be. The universe will recruit you on its own terms, not yours. Meanwhile, your best strategy will be to prepare for this future service as best you can.

The "how?" question (a very brief introduction to group process)

Following your impulse to connect with others who are similarly called, you may also be tempted or persuaded to join or coalesce a group of people working on particular local food issues. This is an appropriate exploration for you, for regional foodsheds can only be built through those joined together in rigorous collaboration and cooperation–skills that have been nearly forgotten in our society. You will experience many levels of working with various groups, and along the way you will discover how the process of cultivating emergence within a group can be nurtured as a shared discipline.

Meanwhile, you will frequently come to grapple with a recurring question that can be a project-killer: *"How are we going to do this?"*

The impulse to immediately plunge into answering this question in the early stages of a project is seemingly natural, but it is *the wrong question at the wrong time.* As an evolutionary foodshed catalyst, you will be learning to *discern deeper currents* in a project so that you can guide such efforts appropriately.

You will learn to discover and articulate the vision that the universe wishes to bring into manifest form. (To speak of what the universe desires is a manner of speaking that points to the long arc of the moral universe, the direction that the arrow of evolution is always seeking.) This may sound metaphysical, but it is profoundly practical, and is largely a matter of seeing and deeply feeling what is truly needed in a situation–as distinct from what is desired by the people involved.

As this vision gains clarity in your own perceptions and those of others, you will learn to activate and mobilize relationships around this vision. And you will learn how to lead a process of discovery among those involved to uncover a thorough understanding of the situation you face together, and the broader context in which this is occurring. You will discover what you must learn together, and what skills and resources you must develop.

From this you will gain an incisive view of current reality, the challenges and obstacles you will need to overcome, and how you as a co-creative group can rise to meet them.

At some point in the process, the overarching question will become, *"Will we do this?"* In other words, knowing what you now know, and understanding who you are together, will you choose as a group to be responsible for shepherding this vision into its full manifestation? When the answer is "Yes!"–a decision and commitment made in the absence of knowing exactly *how* you will accomplish this–then the "how" or underlying design of the project *will be revealed*. It will emerge in unexpected ways.

There is much more to be said later in this book about your role in such groups, and how they can function as foodshed catalysts together, but this is enough for now. If this process seems opaque or confusing to you at the moment, do not be deterred. Start with contemplating what is truly needed in the situation you are facing in your community.

What I am pointing to here is the briefest possible description of how emergence or evolution actually unfolds in a community, and the pivotal role that you may be given to play in this process of constant learning, experimentation, and developing the discipline of being a r/ evolutionary foodshed catalyst. At the moment, you are in the earliest stage of learning this, and you will need to develop patience with yourself and patience with the process. It is mysterious, unpredictable (except in its direction), and impossible to control.

Among the lessons you must learn is that whatever you see is truly needed and wanted, you cannot make it happen. No group can make it happen. The most you can do is to become a vehicle through which the evolutionary healing process unfolds. This is the essence of true r/evolutionary leadership.

It should be said now that there is a great difference between being an activist and serving as an evolutionary catalyst. Activism as it has been known during the Holocene Epoch is a distorted attempt at command-and-control, to make things happen. This is not the way of the evolutionary catalyst, who must learn to align with and embody the evolutionary force itself.

These words may be hard to take in right now, but allow them time to be absorbed into your being. We will come back to this again and again.

Embracing the sacred

Before leaving this chapter, there is one more connection-related issue that is important to acknowledge.

You may already intuit or partially experience this now, but in your life, you are about to discover the meaning of the phrase "Food is sacred."

The reality is that food is our connection to life itself, although in our industrial food system this connection has been almost completely severed–resulting in untold destruction and suffering on this earth.

In discovering the local food experience for yourself, you will come into direct conscious contact with the sacredness of food–local food, of course, produced by those who have surrendered their own personal ambitions in service to the sacred itself.

This is neither an intellectual nor a religious concept, but an experience of how life moves and flourishes on a planet such as Earth. Once we experience the sacredness of food, and how it brings us into connection with what is most holy in life, we can never be cynical or hopeless again. And we can never allow ourselves to be seduced by the industrial food system which seeks to disconnect us and separate us from all that is sacred and holy on this planet.

I need not try to convince you of these things, for you have already tasted them. You already know that food is sacred, and that the growing and sharing of food is sacramental. This is the heart of the agrarian culture, which has largely been stolen from us, ripped away by the greed and ambition of the unholy alliance of big food, big ag, big pharma, and big oil, who–with the support of big banking and big government–have wrested control over almost all of our food supply and have together become the most destructive force in the world.

What you will be coming to terms with in the process laid down in this book is that you have a crucial role with the people of this land rising up: to take back our food supply, and to restore the centrality of sacred food in our society. This is an undertaking of epic proportions, and it will demand of you more than you ever intended to give. But you are destined for this, and at this moment stand on the precipice of moving into the very purpose and meaning of your life.

—

In the next chapter, we will begin to explore together the larger context of our food predicament as part of a shattering planetary crisis.

Fair warning, this material is confrontive and deeply activating. Allow yourself time to take this in, and resolve to receive the wisdom and encouragement that the universe is offering to you now. Know that it is only by confronting the realities of our predicament that we will be able to be truly of service.

As you read on, I offer up my heartfelt prayer for you, that you will receive what is being offered here, and that you will resist the temptation to resist what is reaching out to you.

You have come to this world precisely because of this predicament, and you are greatly needed now. The chapters that follow are meant to prepare you and support you in fulfilling your true purpose.

#

Chapter 5

Current Reality I
(Stage Three)

The Unholy Alliance and the Anthropocene

As foodshed catalysts, we must know what we're facing, the larger context in which we work, the realities of our current situation. Otherwise, we're flying blind.

Opening to current reality is a discipline that we must master, and get better and better at over time. Being in touch with unfolding realities informs the vision that moves us–and that vision will evolve accordingly.

This chapter and the next will explore emerging realities that are rarely discussed in our society, but are crucial for those of us working to build regional foodsheds.

Before you read further, I must be frank: This chapter is a deep dive into the *bad news* of our collective predicament. The news is breathtaking, devastating, and largely unknown in our society.

But the following chapter demonstrates that the coming catastrophe is but an inevitable transition to a way of living on this planet that is the fulfillment of what we deeply long for. So by all means take in the news of this current chapter, but do not stop here. There is another side to the story, revealed in chapter six.

The assessment

As a people, we are just beginning to become aware of the magnitude of the crisis that's rapidly unfolding on this earth. We're still in shock, with most people still in denial. But every day stories appear in media around the world confirming that the future we're heading into is unlike anything we've experienced before. In fact, there's such a flood of new scientific information revealing our predicament that it's almost impossible to keep up. Our scientists are having the same problem.

It's recently become clear that we're experiencing the early stages of the worst global catastrophe since the dinosaurs were wiped out some 66 million years ago–along with 70 percent of all other species of life. The impact of our industrial civilization on the planet's biosphere has become so great that scientists now agree that humanity has become a destructive force of geologic scale and that our activities in recent centuries are forcing the entire planet into a new planetary era, the Anthropocene Epoch–marked by a mass extinction of species, only the sixth in earth's 4.5-billion-year history. Some 100 to 200 species are now disappearing every day.

Global warming is the primary engine of the Anthropocene, with greenhouse gas emissions–primarily CO_2 and methane–driving up temperatures far more quickly than almost any climate model had predicted.

Unexpectedly, excess carbon dioxide in the atmosphere (now at about 405 parts per million, the highest level in at least three million years) is disrupting the photosynthetic process in plants–especially grains and cereal crops–sharply reducing their nutrient value to animals (including humans), and damaging their ability to support healthy soils.

Shockingly, each degree Celsius of temperature increase will reduce global food production capacity by at least ten percent. Since an increasing number of climate scientists now quietly project an increase of 4 - 6 degrees C by 2050, it appears likely that global food production will be cut in half *over the next 32 growing seasons.* Inevitably, famine will soon be widespread.

Peter Wadhams at the University of Cambridge brings the truth home: "Sooner or later, there will be an unbridgeable gulf between global food needs and our capacity to grow food in an unstable climate. *Inevitably, starvation will reduce the world's population.*"

Over the next hundred years or so, human population will indeed rapidly decline–to perhaps only one to two billion people. Starvation, disease, war, and catastrophic weather events will be the primary factors in this decimation.

Abrupt climate change is a current reality, and the devastation it heralds is beyond the capacity of most of us to contemplate.

Welcome to the Anthropocene, the age when planet earth plunges into darkness and great uncertainty.

For those of us who are awake and aware, all this is deeply unsettling. But, as Ralph Waldo Emerson suggested, only to the degree that we are unsettled is there any hope for us.

The indictment

Here comes more unsettling.

We now know that the primary cause of this global disaster is totalitarian agriculture–the way humanity has organized to feed itself–which has now been industrialized and globalized by an unholy alliance of big ag, big food, big pharma, and big oil, and empowered by big banking and big government. In the last hundred years, the global industrial food system has become the largest and most destructive industry in the world, the primary contributor to global warming.

A hundred years ago our entire food supply was local and organic. Today, 99 percent of our food supply is industrial–non-local (imported), and non-organic–and controlled by the unholy alliance. This is the result of the most massive and most successful colonization effort in human history.

This industrial food system is destroying life on the planet, and it is killing us. Of all the people who die each year in the U.S. and Europe, approximately half now die from food-related illnesses–a direct result of the industrial food system. The toxins, chemicals, hormones, and poisons in industrial food are unleashing an epidemic of disease unlike anything in human history.

While the so-called Green Revolution–the displacement of agrarianism in third world nations by industrial agriculture–caused a tremendous spike in human population, to the point of population overshoot, the backlash is coming due as global warming causes the collapse of the industrial food system and triggers the collapse of human numbers on the planet.

A matter of knowing

Perhaps it is occurring to you by now that I am not providing persuasive arguments and evidence for these realities. This is because I am not attempting to persuade you of anything here. These are matters that with a bit of research you can easily confirm for yourself. Once you begin looking, you will find the corroborating evidence everywhere. My intention here is to share with you the unvarnished realities as we have come to know them, the facts you need to guide your efforts as a foodshed catalyst.

Yes, all this is deeply unsettling. But it is essential to know and understand these things. As environmental lawyer and advocate Gus Speth says, "Unless we know the full extent of our predicament, we will never do what is necessary."

For the communication in this chapter to be of any lasting value to you, you will need to take your own journey to confirm whether this assessment of current reality is true. Do not shortcut this process. Until this becomes a reality for you, you will remain confused and ambivalent.

It is important to say that it *is* possible for you to actually know what reality is. Humans have the innate capacity to know, and we each need to develop this. We do not need to settle for merely believing. We fear that we can never really know the truth, and therefore compromise by adopting comforting or even self-destructive beliefs. But we do have the inherent ability to know, and our deepest feelings and sensibilities can be our guides (if we learn how to access them).

Living in a dying civilization

We live in a society where most people are unaware of the full extent of our predicament, often by choice, often choosing to believe something else. This is typical in a dying civilization.

Yes, our civilization is dying, as all human civilizations have.

The genesis of our current global industrial civilization is in the practice of totalitarian agriculture, which began about 10,000 years ago. This was a dramatic departure from the bioregionalism of indigenous peoples, and it formed the foundation of an industrial civilization that has come to dominate the planet—the first global civilization in planetary history, almost entirely displacing all other cultures.

Now, the industrial food system (and its resultant global economy) has grown to such a scale that it is undermining the planet's ability to support life. This is an unprecedented global disaster.

We are learning it is no coincidence that every civilization has collapsed because of an agricultural disaster, as ours is beginning to collapse now.

The usual automatic response to the reality that our civilization is collapsing is denial, for our limiting beliefs make it impossible to envision the inevitable.

You, however, do not have the luxury of such denial. You already know these things in your heart, even if you may not have all the corroborating evidence yet. You can see and feel the full extent of our predicament. And you have the growing sense that you have some role to play in the unfolding drama of the coming collapse.

The shock of recognition

About ten years ago, I had an opportunity to speak publicly about these things at an unusual conference in Denver, and was even featured in a dual presentation with Richard Heinberg, Senior Fellow at Post Carbon Institute, lovingly nicknamed "Dr. Doom" by his friends.

Unbeknownst to me, in the audience was Dana Miller, a recently retired airline hostess. A cheerleader since adolescence, "pompom Dana" was unprepared for what Heinberg and I were saying about the full extent of our predicament, and she was devastated. As she tells it, she went home from that conference and cried for three days.

Then she went to work, soon becoming a key leader in Denver's local food movement, pouring all her energy and enthusiasm into this urgent cause. I first met her many months later.

Dana is a r/evolutionary foodshed catalyst, and is one of my heroes. She allowed herself to come to grips with current reality, and in response determined to give her life in service. But it was not argument or evidence that persuaded her. What was activated was her own deep inner knowing.

You need not attempt to persuade people of our collective predicament, and you certainly should refrain from argument. Presenting endless confirming information will have little effect. Information does not change people's minds or hearts. You must instead speak to the part of people that already knows the truth. This kind of communication relies on transmission rather than persuasion, and it has the power to activate or ignite what has been hidden deep within the individual.

The update

My own comprehension of current reality was shattered once again in early 2016, when I stumbled upon a video of a long presentation given by Dr. David Battisti at a climate conference hosted by the University of Arizona in Tucson. Battisti is the distinguished chair of the Department of Atmospheric Sciences at the University of Washington.

The audience seemed to primarily consist of scientists and climate activists, most of them very sophisticated and very well informed. (When I say I "stumbled" upon this, I should explain that it is my habit to constantly search for what's new and significant in human understanding of our current reality, as well as the direction and process of evolution itself. Every day I quickly scan numerous publications, blogs, book publishers, and social media, and capture what seems most important. For me, this discipline is but a part of a commitment to perpetual learning. But I'm not looking for information so much as I am seeking that which activates what I most deeply already know.)

The title of Battisti's talk immediately caught my attention, "Climate Change and Global Food Security." I knew of no other climate scientist who was exploring the relationship between global warming and food production. I found it promising to see that Battisti was suggesting this was a matter of *food security*. I was even more intrigued when I learned that he had grown up on a dairy farm in upstate New York.

Battisti's core assertion is that heat stress alone reduces crop yield by roughly ten percent per each degree Celsius. He takes time to lay out plenty of evidence to back up this assertion, and shows how the breadbasket regions of the world are among the most temperature-sensitive.

To make projections into the future, Battisti chooses to base them on the United Nations IPCC scenarios. From these numbers, Battisti is able to reasonably declare that global food production capacity will be reduced by 30 to 40 percent by the end of the century.

This is breathtaking, an unmitigated catastrophe.

But as I listened carefully to Battisti, I realized that he is actually playing it safe and giving us a very optimistic *best-case* scenario.

As you probably realize, IPCC data are notoriously outdated and incomplete, producing an unrealistically conservative view of our climate predicament. Battisti skirts this issue, but proceeds to reveal that temperature increase carries with it related effects that will significantly impact food production.

For instance, he says that increased pest pressure alone could cause an *additional* ten percent reduction in crop production per degree of warming. He goes on to quickly discuss other factors, including increased disease transmission rates, changes in rainfall patterns, droughts and decreased water supplies, and catastrophic weather events–plus loss of land due to soil erosion, desertification, salinization, and sea level rise–plus the reduced nutrient and protein levels in crops caused by excess CO_2 (as recently documented in a groundbreaking study by Dr. Irakli Loladze).

Being a good conservative scientist, Battisti refrains from making projections beyond what the ultraconservative IPCC data would support. He leaves it to us to connect the dots and apply the principles he has laid down.

So what are we to make of all this? I consider IPCC future scenarios to be wildly unrealistic and overly optimistic. I have learned of an increasing number of climate scientists who are quietly indicating that increases of four to six degrees Celsius by 2050 are now almost inevitable, perhaps reaching 10 to 14 degrees by the end of the century. When I listen to them, I know that they are far closer to the truth than the IPCC.

For instance, Dr. Kevin Anderson of the highly-respected Tyndall Center for Climate Change Research in the UK said in 2012 that we are almost *guaranteed* to go to *at least four degrees Celsius warming—* and perhaps even more than that—*by 2050.*

No projections are certain, and all this is impossible to quantify, but we can easily know that a radical change in global food production capacity is coming in the very near future. We are facing abrupt climate change and a global food emergency. Widespread famine is inevitable.

This is nothing short of a global food catastrophe. We have no more than a few years to prepare. The only way to prepare is by building lifeboats, highly localized and deeply resilient regional food systems. This must become the highest and most urgent priority of our communities. There is nothing else that can prepare us for what is coming.

Global warming is upon us, and its devastation will be far worse than we've been told. Yes, even right here in the U.S.

Anyone looking at such realities could only conclude that we're in for some terrible shocks, and a global food crisis of historically unprecedented scale. Unfortunately, almost no one is looking. This should be front-page news around the world every single day.

But for those of us who are looking, it's not difficult to understand that we're headed towards a global food catastrophe as the global industrial food system begins to collapse under the strain of global warming.

We can see where this is going. We are plunging into a crisis for which no one is prepared, or even preparing. Actually, "crisis," is too tame a word. We could just call it "the Anthropocene."

Confirmation

I have shared this perspective with many audiences now, and I can say that almost no one among them has been aware of Battisti's presentation. For me, what he is pointing to changes everything for those of us involved in developing local food systems.

A strong clue is his suggestion that the areas hardest hit will be the grain-producing breadbasket regions of the world, including the Midwest of the U.S. These are the areas where industrial agriculture is most deeply entrenched.

Lest this seem too radical an idea, little-known earlier studies have previously pointed in this direction, though not with the scientific detail that Battisti provides.

A high-level 2014 report on the risks that climate change poses to agriculture in the U.S. warned, "Under the 'business as usual' scenario and assuming no significant adaptation by farmers... the Midwest region as a whole faces likely yield declines of up to 19 percent by midcentury and 63 percent by the end of the century." This prescient report was co-authored by none other than Henry Paulson (who was George W. Bush's Treasury Secretary) and George Schultz

(Ronald Reagan's Secretary of State). We can safely assume that the conclusions of this report are rather conservative. Unfortunately, in today's political climate such perspective is widely ignored, and this report has essentially been buried.

The World Bank commissioned a study nearly a decade ago (subsequently suppressed), which concluded that if global temperatures should increase by just two degrees Celsius above baseline, India would lose 25 percent of its food production and China would lose an astonishing 38 percent. This helps us understand that China–who, with 1.4 billion people and more than 25 percent of its agricultural land doomed to desertification, understands the problem very well–is now desperately buying up huge tracts of farmland in Africa and South America so they can feed their people. "China is positioning itself for the struggle to come, the struggle to find enough to eat," says Peter Wadhams at the University of Cambridge. "By controlling land in other countries, they will control those countries' food supply."

Mobilizing the revolution

It is obvious now that the concentrated industrial-scale monoculture practices of big ag will be unable to quickly adapt to the rapid increase of temperature and its attendant effects. Industrial agriculture will be hit hard, to the point of collapse, unleashing a global agricultural disaster–a fitting end for a dying civilization that was birthed with the advent of totalitarian agriculture.

Here's the bottom line: Since in this country we are dependent for more than 99 percent of our food supply on the unholy alliance's industrial food system, we must quickly mobilize to be able to meet our communities' food needs as locally as possible.

Therefore, *we must declare a food emergency now*. The lives of our children and our grandchildren depend on us doing this.

This is why we are calling for a local food revolution–a radical expansion and acceleration of the local food movement in impact, effectiveness, and scale.

This must be the equivalent of a wartime mobilization effort. Localizing our food supply must become the most urgent priority in every community. Failure is not an option.

But who will mobilize this herculean effort?

The reality is that there is no one to do this in your community but you. You cannot look to anyone else to do this. This falls to you–and each of us who comes to grasp these realities.

If you can begin to see this and feel this, then you will realize that we have no choice but to rise up as a people and take back our food supply.

The problem is that today almost no one understands *why* this is necessary, for this has all been hidden from us.

Who will tell the people in your community that we must shift into emergency mode to build a highly-localized regional food system, a living foodshed? Who will sound the alarm? Who will tell them about the devastating news from Dr. David Battisti, and Dr. Irakli Loladze? Who will show them how this can be done? And who will be able to stand up against the inevitable denial?

This comes down to you. There is no one else to do this.

—

Once I understood the implications of the work of Battisti and Loladze, I realized that there was no one in Colorado even beginning to prepare for the future that is headed our way. Almost no one suspected the connection between global warming and food production capacity, except that some thought the carbon sequestration benefits of organic food production could eventually reduce CO2 emissions.

This meant that it came down to our organization, and a handful of others working with us, to begin to shift the entire conversation about local food and to mobilize an appropriate response. We bore the responsibility for this.

—

Meanwhile, the unholy alliance–sensing last-chance opportunity– is committed to wringing out the very last profits possible and seizing whatever power they can–however temporary–before their own ignominious collapse. Their demise will be spectacular, and it will come soon, and the disruption to the industrial food supply will find many nations completely unprepared–maybe even especially the U.S., which has earned a reputation as the world capital of denial.

So, given all this, it's clear that our only responsible choice now is to immediately shift into *emergency mode* to organize our communities to feed ourselves and each other as locally as possible.

We must reclaim our food sovereignty and establish our food security as quickly as possible, as locally as possible. This will require a revolution–a *local food revolution*!

This means that you must not only become a revolutionary, but you must begin to lead the local food revolution in your community.

#

Chapter 6

Current Reality II (Stage Three)

Emergence and the evolutionary perspective

You've probably had enough difficult news for now. The material in chapter five can be very hard to take at first, because it makes undeniably obvious that localizing the food supply is an urgent necessity in the face of unstoppable global crises.

Unstoppable? Well, yes. The genie is out of the bottle, so to speak, and we can't push it back in. Pandora's Box has been opened. We've gone past the tipping point. We've crossed over the line into the Anthropocene. We've unleashed a famine of epic proportions, along with a planetary extinction of species (possibly including our own).

Seeing this and feeling all this can be devastating, and so it is for many people–even though most are not able to consciously identify the source of their pain and angst.

In this chapter, you will begin to discover that this convergence of global catastrophe which we call the Anthropocene Epoch is actually the beginning of an *evolutionary course-correction* now taking hold on this beleaguered earth.

Once this sinks in, this will come as profoundly good news, the completely unexpected *other side* of our collective predicament. But this is nearly impossible to know without some sense of how evolution–and the process of healing and regeneration–actually works.

Also in this chapter, you'll begin to see and feel how highly localized regional foodsheds are the gateway to this evolutionary course-correction process, and how you can play a crucial role in this great coordination.

The evolution of evolution

I don't know any reasonable way to introduce this new material except through a very personal story.

For years I have been struck by the almost mystical pronouncement of Martin Luther King Jr., in Selma, Alabama nearly 50 years ago as he led the battle against racism segregation: "The arc of the moral universe is long, but it bends towards justice."

These are prophetic words of the deepest sort, and only in recent years has it been possible to clothe them with grounded meaning.

Such understanding first began to appear in the writings of Pierre Teilhard de Chardin, a silenced Jesuit priest and paleontologist of the early 20th century whose works were banned by the Catholic Church until after his death in 1955. It's possible that King knew of Teilhard's views, but I feel it's significant that the sense of "the arc of the moral universe" was already finding expression in modern culture.

Teilhard de Chardin was among the first to see that the universe has been *evolving* from its beginning and that it has a definite direction—in his cosmology, towards a reunion with the divine.

Teilhard had an inexplicable impact on my own life, for when I was a senior in high school I somehow came into possession of one of his previously banned books, *The Future of Man*, which was my first glimpse of the evolutionary trajectory that humanity is following, from geosphere, to biosphere, to noosphere, to Point Omega. (How did this book find me in a little town of some 2,000 people on the plains of northeastern Colorado?)

Some three decades later, an uncensored Catholic priest–cosmologist and "geologian" Thomas Berry—introduced physicist and mathematical evolutionary cosmologist Brian Swimme to Teilhard's work. The collaboration between Swimme and Berry resulted in what is likely one of the most important books of the 20th century, *The Universe Story: From the Primordial Flaring Forth to the Ecozoic Era*, which drew upon the latest scientific understanding to trace the evolutionary arc of the universe from the Big Bang to the present. For the first time, we could see the unfolding of the physical universe *as a story*—with direction, meaning, and even purpose.

Swimme and Berry wove together, in a most beautiful way, the story of how our universe developed. It was essentially a history. My sense is that finally being able to describe the unfolding of the universe as a coherent story should be considered one of the great achievements in human history (a catalytic event, really), one that could not have been possible before the closing years of the 20th century.

This was especially meaningful for me, because a decade or so before *The Universe Story* was published, I had the life-changing opportunity to study the work of Arthur M. Young, an American cosmologist, philosopher, and inventor of the first commercially successful helicopter (the Bell Model 30, prominently featured in the television series M.A.S.H.). In 1976, Young published the pioneering book, *The Reflexive Universe: Evolution of Consciousness*, the first scientifically-grounded attempt to show that the evolution of the universe proceeds in a pattern of seven distinct stages of increasing complexity and wholeness, and that this same pattern is embedded in all physical, biological, social, and spiritual processes. I was learning how to apply this process in organizational development and the creation of transformational communication.

With *The Universe Story*, I could begin to fill in the details of the seven-stage evolutionary process that Young so presciently identified, and the model held true. Now I had a tool I could use more precisely. And when Lynette Marie and I met some 21 years ago, I spent the first months sharing with her everything I had learned about the evolutionary process. None of this was a surprise to her, thankfully, and we embarked on this evolutionary journey together. She had actually been using a very similar seven-stage process in her mediation and conflict transformation work, but didn't realize it was tied to the process of evolution.

Teilhard, Young, Berry, and Swimme have in effect been my teachers and guides for much of my life. But they have not been teaching me so much as they have been activating what I somehow already knew in my soul. They gave me the confidence to follow the evolutionary arc of the moral universe, and to learn to give it expression in new ways.

Another important character in the drama of humans coming to understand the nature of evolution is South African Jan Christian Smuts, whose 1926 book, *Holism and Evolution*, and subsequent teachings have been foundational to Zimbabwean Allan Savory, the founder of Savory Institute (based in Boulder), and the creator of Holistic Management, a systems-thinking method of managing the relationship between humans, livestock, and land to reverse desertification and end global warming. So Jan Smuts and Allan Savory are implicated in the evolving story of humans coming to grips with the principles of evolution.

The other side of current reality

Well, enough backstory. This is really a way of introducing the other side of our current reality–the reality of how evolution works and where we are in the process.

What we have learned is that evolution unfolds through *emergence*. That is, the process of evolution is self-organized, and consistently reveals itself in patterns of increasing complexity and wholeness–a cycle of seven radical transformations or stages that each initiates fundamentally new capacities, each incorporating the properties of all previous stages.

The following diagram shows the basic sequence of the process of evolution: orders of existence that increase in complexity, each stage built upon the foundation of everything that went before.

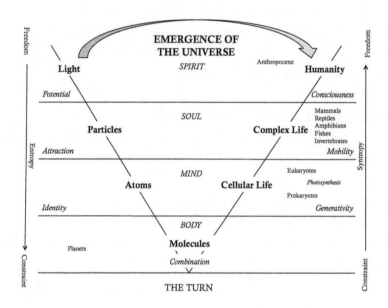

We've learned that the universe does not merely evolve. The universe itself is a living process of emergence, which moves progressively through patterns of radical self-transformation which are mirrored at all levels of existence, from the macro to the micro.

What becomes clear is that the universe is self-organizing, self-transforming, self-healing, is progressive and directional, reflecting a greater force or organizing principle that is being expressed through its unfolding.

We have also learned that it is possible to not only learn these patterns, but to consciously embody them and to apply them to human endeavor.

If you know where you are in an unfolding evolutionary process, you not only know where the process is going but what you need to focus on next–and what the challenges and opportunities will be.

For instance, I've already said that this global industrial human civilization is dying. That may seem pretty horrifying–unless you understand that death and seemingly chaotic cataclysm are characteristic of a mid-point in the evolutionary process that gives rise to *emergence*. Out of the ashes of a dying civilization, something new emerges. Out of the ashes of a mass extinction event, the way is cleared for new and more complex life forms (98 percent of all species that have lived on this planet are now extinct). Out of the ashes of a dying star (i.e., a supernova explosion), more complex molecules are generated that ultimately make life possible.

Out of the ashes of a dying relationship, something new is born.

Out of the ashes of a dying industrial food system, a network of new and highly localized regional foodsheds emerges.

Emergence follows cataclysm. When the evolutionary process reaches impediments or obstacles, cataclysm clears the way for emergence. This cycle is fundamental to all life.

Emergence is the most important framework for the evolutionary catalyst, for we are not seeking to create change. We are becoming the vehicles through which emergence unfolds. We are, in effect, manifesting the fundamental evolutionary force. We serve not our own will, but that of emergence. You could say that this entire book is about emergence.

How we got here

Perhaps it's now useful to very briefly share what the trajectory of modern civilization looks like from an evolutionary perspective.

This civilization began with the appearance of a particular form of agriculture that sought to feed growing human populations in villages, towns, and cities by essentially clearing nearby land of all living organisms in order to produce a limited number of food crops (animal or vegetable) at a sufficient level of scale that ensured enough surplus to support the population beyond growing seasons.

As we have seen through time, this totalitarian agriculture depends on great violence and cruelty to advance its aims. It is also profoundly unsustainable. But it provides fertile ground for those who seek power and wealth. In fact, it has provided the very origins of our global economy.

Over time, the process of increasing urbanization of the human population and the destruction of natural habitats to produce food for humans–and the struggle for control over such systems–has greatly expanded, and this has led to industrialized food production on a global scale. Along the way, food has become commoditized and trivialized–and desacralized.

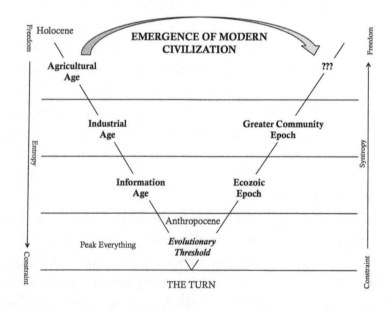

It's possible to chart the trajectory of modern civilization in evolutionary terms:

Age of Agriculture (stage one): As I've said, what we think of as human history actually began with the birth of the age of agriculture–totalitarian agriculture (or agricultural colonization). But this is just a tiny slice of human history, and is actually the history of a civilization–our modern civilization–that ran its course over 12,000 years. Before the age of agriculture, humans did not have mathematics, or written language, or cities. Written language only evolved after the so-called agricultural revolution, first in the form of numbers. In fact, agriculture birthed economics and finances–the language of numbers–long before language of expression came into being. So it's extraordinarily difficult to discern human history before this time.

Age of Industry (stage two): Once totalitarian agriculture got under way, it remained the dominant organizing principle of this civilization until the 1700s, when a revolution in economic and social organization brought about the age of industry. The industrial revolution represented a shift from relying on human muscle power to the awesome power of machines driven by fossil fuels.

Age of Information (stage three): The next stage of evolution of modern civilization came during the late phase of World War II, as the emergence of digital technology ushered in the information age. Both agriculture and industry have been greatly enhanced by the technology of the information age.

These first three stages of the emergence of modern civilization–the age of agriculture, the age of industry, and the age of information–are cumulative, building on each other, and they overlap. All three are in full bloom today.

There is nothing linear about this process. There are feedback loops between the stages. But what is harder to see is that the entire process is moving in a direction, and with increasing complexity.

The Turn (stage four): It's also worth noting that through these first four stages, the evolution of human civilization seems to be greatly accelerating, to the point that we're already being swept into the next stage, which is a turning point in the evolutionary process. Cumulatively, the developments of the first three stages of the evolution of modern civilization have brought us to the brink of global disaster and apparently irreversible entropy. They have also delivered us into the belly of an evolutionary threshold or bottleneck–thrusting us into a new geological era or epoch or age, the Anthropocene. (In geological terms, an age is millions of years, an epoch is hundreds of millions years, and an era is several hundred million years. For now, no one knows how long the Anthropocene might last.)

A pattern begins to emerge here. In its early stages, evolution moves from freedom to constraint, with increasing entropy. Finally, in stage four, constraint and entropy become extreme, opening up a time of crisis and chaos, and descent into seemingly irreversible entropy and inevitable extinction. But this is only part of the story, only one-half of the arc of evolution.

Ecozoic Epoch (stage five): Thomas Berry and Brian Swimme have called the stage following the Anthropocene the Ecozoic epoch, an entirely new phase, in which humanity not only devotes itself to repairing the destruction it has inflicted upon the world but also engages in a new relationship with nature and with life itself. Here, with the partnership between humanity and the community of life, the biosphere gradually becomes self-renewing and ecologically sustainable.

Greater Community Epoch (stage six): If humanity can attain a significant level of equilibrium in the Ecozoic, we may then emerge into a realm of existence that goes far beyond our planetary sphere, perhaps engaging in a greater community of intelligent life.

Stage seven: Our ultimate destiny is unknown and unknowable at this point. But it will likely go beyond (and yet still include) the physical universe.

The emergence of foodsheds

Our first book, *The Local Food Revolution,* is partly an attempt to apply the principles and patterns of evolution to the process of food localization. Our fundamental discovery is that food localization is a natural manifestation of evolution itself, and represents the emergence of healing and regeneration on planet earth. In that book, we laid out the first steps towards "a pattern language" for food localization: an identification of the sequence and patterns of relationships by which this process naturally unfolds.

If you are familiar with permaculture (which is a practice based on an understanding of how life works on this planet), you may be familiar with *pattern language,* which originated with legendary architect and evolutionary philosopher Christopher Alexander, first formally expressed in his 1977 book, *A Pattern Language: Towns, Buildings, Construction*–which became the inspiration not only for permaculturists but agile software developers and social activists alike.

Here's a glimpse of the broad patterns of the emergence of a highly localized regional foodshed (we'll explore this in more detail in chapter eight).

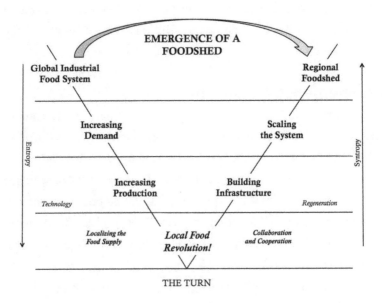

What you need to know now

But this book is not about understanding the evolutionary process. Instead, it is meant to give you a new perspective, to raise the possibility that food localization is right at the front lines of evolutionary course-correction. You are a vehicle for this evolutionary course-correction.

At this point, it's not necessary for you to understand the details of how this works. What's necessary is to understand where we are and where we're going, and to have a general sense of how we're going to get from here to there.

You can feel, if you allow yourself, where we need to be going. You can sense the direction of the long arc of the moral universe, where the arrow of evolution is attempting to take us.

Evolution, as Martin Luther King, Jr. reminded us, bends towards justice–relentlessly. The flight of evolution's arrow is guided by what we would have to call love. Evolution itself–the entire process–is love in action. Love manifests through emergence. You can feel this.

In the course of the evolution of a galaxy or a planet or a species of life, things can and will go wrong–temporarily. Things can fail–from the micro to the macro–temporarily. But the evolutionary impulse cannot be thwarted, and will always find ways to correct course and break through. The direction is constant. The patterns of relationship and process are persistent. The process is relentless. Therefore:

1. What you first need to know is that where the evolutionary process is taking us is to precisely what you most deeply want and need (which, thankfully, are identical).
2. What you need to know next is that you can trust this process, which will proceed with healing, restoration, and regeneration.
3. And next you need to know that you can learn to align with and even embody this evolutionary process, to consciously become its vehicle, its agent–to the point that you become the very place where the dream of the universe and the dream of the earth and the dream of humanity all meet, converge, and manifest into multidimensional reality, from the physical to the spiritual.
4. Finally, you need to know that all humans have these capacities.

At first, these appear only as possibilities, but they represent the *promise* of evolution. You are invited to trust this promise, until you evolve to the state where you can actually know this and live this.

Now I am speaking of the path of the evolutionary catalyst, one of learning to serve as a vehicle for evolution itself. Yes, this is a spiritual path, and it can be learned and practiced. But that is advanced work, not for this book.

For now, you can learn to discern, align with, and support the ways evolution seeks to unfold in humanity's relationship to food and agriculture. This book is an introduction to serving as an evolutionary foodshed catalyst.

What you will discover is that the historic trajectory of our current civilization is fundamentally out of alignment with the direction of evolution. We have been out of sync with nature since the birth of totalitarian agriculture, which is where we took a wrong turn.

The consequences have eventuated in a planetary crisis, a cataclysm which we have come to call the Anthropocene. But out of cataclysm, emergence always arises as a fundamental evolutionary course-correction.

As a people, we have finally arrived at an evolutionary turning point, where the cataclysm we have precipitated begins to give way to emergence, the reversal of entropy, the return to the unfolding of evolution—out of the ashes of the dying civilization, out of the ashes of the dying industrial food system.

In human terms, this course-correction will unfold over a very long time—even though the situation is urgent—but it will be very thorough and absolutely guided by love (which, it turns out, is the underlying design of the entire universe and of its evolutionary process; in other words, as King himself implied, the universe is not only moral and just, but *loving*).

In evolutionary terms, this course-correction will unfold quite rapidly. The transitional period of the Anthropocene may last for a few centuries or a few millennia. How long it takes does not seem to matter. That is, how long it takes is just how long it takes.

Evolution is relentless, and will always ultimately find the way to manifest what is needed, that which is the expression of the underlying design–whether in a galaxy, or a planet, or an intelligent species, or a foodshed. You can rely upon this process, and you can rely on your capacity to be a conscious part of it.

This is the good news that is only beginning to be discovered in the 21st century. This understanding, rather colloquially known as "the evolutionary perspective," represents what we could call a quantum leap in the evolution of human consciousness. Over time, the ripple effects will be far reaching, and you will be part of it.

The process of evolution is a discovery as sweepingly profound as the discovery that our earth is a planet orbiting a star in a galaxy of billions of such stars in a universe of trillions of such galaxies.

We cannot yet imagine what will be unleashed in this universe as a result of the evolutionary breakthrough that is beginning to occur on our planet and in us as a people. But it is becoming apparent that we have vastly underestimated our potential as a species of intelligent life–a potential we may begin to discover and even experience if we are able to rise to the occasion and successfully navigate the treacherous seas of the Anthropocene.

If we are successful in this endeavor, it will be in no small measure because we have learned to build and thrive in the "lifeboats" of regional foodsheds.

And it will be in no small measure because we have learned to consciously allow the process of evolution to move in us and through us.

This is the destiny that we will share together.

#

Chapter 7

Radicalization
(Stage Four, The Turn)

This chapter is going to be different from the rest.

This chapter is the turning point of this book, and the turning point for you. It is here that you will begin to make the decisions that will determine whether you will be able to align with the evolutionary process and lead the local food revolution in your community. *Are you up for the challenge?*

This gets rather personal, so perhaps it's best if I first say something about the process of radicalization I've been going through myself over the last few years.

For some time, I had been feeling that the local food movement was somehow failing. Despite all the progress that had been made around local food in Colorado, despite all the very exciting food and farming projects that were showing up through the state, I could see that we were a long way from communities actually working together to build highly-localized regional food systems resilient enough to thrive during the uncertain and challenging times ahead.

It was painful to me that most people seemed to view local food as an optional and expensive lifestyle choice for the privileged, and that they did not recognize that food localization was needed as a matter of food sovereignty and food security. Almost no one was thinking or seeing *systemically*. As a result, mobilizing broad community support or significant amounts of capital for food localization was maddeningly difficult. I finally had to admit that I had found almost no one else in Colorado who was thinking in terms of localizing the food supply.

I found it dismaying that Colorado had slipped to #33 on the Locavore Index of states, and that despite all our efforts we were still at no more than one or two percent local food. I learned that few states were doing much better, with the exception of Vermont (where concerted efforts over the last 15 years have managed to bring their food supply up to around eight or nine percent local). When I confirmed that local food represents only 0.3 percent of the total $5.3 trillion U.S. food and beverage industry, I began to see that food localization had a very long way to go indeed. We had been naive and wildly unrealistic in our hopes and expectations.

Around this time, I began feeling that the local food movement needed to go through a radical transformation–an unprecedented acceleration and expansion in impact, effectiveness, and scale. I began calling for a local food *revolution*, and wrote a rather lengthy book partly to discover exactly what that might mean. My public presentations became increasingly strident. I felt pushed to say what no one else seemed willing to say (or even see), and I was surprised how difficult it was even for people involved in the local food movement to take in what I was saying to them. I felt that I might as well have been speaking in ancient Greek.

What I understand now is that I was in the process of becoming radicalized, driven to significantly increase my own level of commitment and impact. I was preparing for the next stage of my work.

Over the edge

More than anything else, what finally pushed me over the edge was watching climatologist David Battisti patiently describe how global warming reduces food production capacity by ten percent for each degree Celsius of temperature increase–and then calculating for myself that, all things considered, global warming was likely to cut food production capacity in half by 2050. With this better-informed context, I knew that we had to quickly find a way to take our local food work to a whole new level. I knew that as a people we were in far greater trouble than almost anyone realized, and that we would have to hurry.

I came to understand that for many reasons–with global warming at the top of the list–localizing our food supply is the most important and most urgent issue of our time. After all, it wasn't raining when Noah built the ark.

I had long known that the unholy alliance's industrial food system had stolen our food supply, and had stolen our democracy as well–by design–and that this was the greatest colonization effort in human history.

But now, with the understanding that the unholy alliance was the primary cause of global warming, it was becoming all too clear that they were also stealing our future right before our eyes. This led me to finally begin speaking publicly in a new way:

Waking up to all this should be enough to set us marching in the streets in absolute revolt. It should be enough to cause riots in every major city across the land. It should be enough to have us burn this system down to the ground. *Desperate rage* is too weak a term to express what we should be feeling.

And that's exactly what the Arab Spring Rebellion was all about, what the people there were feeling, and why what's happened in Syria is so important for us to understand.

We are waking up to the realization that taking back our food supply, taking back our food sovereignty, is the most important and most urgent issue on the planet today, that this is how we reclaim our democracy–and how we get our future back.

But a raging revolt is not what is needed now. What's necessary is a *revolution*, a revolution unlike anything ever seen before in human history. What's needed, and what's coming, is a revolution for healing, for restoration, for regeneration. This is inevitable, and it will be unstoppable.

Never before in human history have we seen a regenerative revolution. But that's precisely the kind of revolution we seek to catalyze. It is not violent or even confrontational, but it is deeply subversive.

The seeds of this revolution are already among us. Now they must be planted and nurtured. This is going to be a very long-term and sustained revolution. But we will need to hurry, for the season's window of time for planting seeds is quickly narrowing.

For nearly three years I had been speaking about the need for a local food revolution. But I can see now that I was doing this as if I was expecting *someone else* to organize and lead the revolution. Somehow it had escaped me that my own radical transformation was taking me to the point where I would be willing to take responsibility for actually *leading* the local food revolution.

I can tell you that this is a very uncomfortable realization for a reclusive writer!

As a result, Lynette Marie and I finally felt compelled to convene the Local Food Academy, a three-month intensive online boot camp for emerging foodshed catalysts who will be in a position to lead the local food revolution in their communities. This is where we are putting all of our energies and attention now.

In the crosshairs

But in this book, it's *you* "in the crosshairs." It is you who are being called to allow yourself to become radicalized into a leadership role in the local food revolution.

Shortly after the American Revolution was underway, Thomas Paine wrote a series of pamphlets titled "The American Crisis." He began with these famous words:

"These are the times that try men's souls. The summer soldier and the sunshine patriot will, in this crisis, shrink from the service of their country; but he that stands by it now, deserves the love and thanks of man and woman."

The "summer soldier" refers to members of the Colonial Army who quietly went AWOL during the harsh winter months. The "sunshine patriot" was Paine's term for those who preferred a more *convenient* war that could be fought quickly with little or no personal discomfort.

We know all too well the summer soldiers and the sunshine patriots of the local food movement, who frequent the farmers markets and upscale restaurants in search of local food, or who blissfully toil in their gardens. Bless them, for they too are needed.

But you are called to be something different. The question is whether you will yield to what moves in you and seeks expression through you. Will you accept the calling to rise to a level of true revolutionary leadership in your community, to spearhead the effort to build a resilient regional foodshed?

—

Perhaps by now it is sinking in that the time you're living in is a pivotal moment in the evolution of humanity and the evolution of the planet.

Perhaps it's become undeniably clear to you that the earth, or nature–evolution itself–is initiating a healing process on this planet and inviting you to become its agent. Others may say that God or Great Spirit is initiating this process. Either way, the process is the same–and it needs you.

You know that the damage to life on this planet is very great, to the point that we're facing the sixth mass extinction of species, the greatest planetary biological catastrophe since the dinosaurs–along with 70 percent of all other species–were wiped out 66 million years ago. Even our own species is in danger of extinction now.

At this point, it's still possible to reverse this damage. It will take a very long time, of course, and the window of time in which we can make the crucial difference is rapidly shrinking. What you do *in the next five to ten years* will profoundly affect the nature of life on this planet in the millennia which follow. Whether humanity will even have a viable future will be determined in this very narrow window of time.

This is a time of great peril and great promise. You now can see that where the evolutionary process of healing and regeneration is beginning to emerge on this planet is around food and agriculture. It's in the local food movement that you find the beauty of the healing process springing up out of the ground, spontaneously emerging in people's hearts and minds.

The ground has been prepared. You are being called. The spirit of healing is moving in you and through you.

It is a planetary crisis around agriculture and food that calls you– in fact, *a global emergency.*

Those of us who see this and feel this often feel almost completely alone. And what this means is that we now bear great responsibility to respond with the very best and the most we have to give. *This is it.*

This next five to ten years represents the most crucial moment in all of human history. This is the pivot point. If we are successful, perhaps no one will remember what we are about to do. If we are not successful, perhaps there will be no one left to remember.

Saying yes

Saying yes to this evolutionary impulse is not a one-time choice, but a daily discipline. This is a practice, even a spiritual practice. There are no proscribed formulas nor rituals, no directions.

This is something you allow to be generated within yourself, and you must learn to trust your gentle internal nudgings as the strongest guidance you may ever receive.

You must learn that you can trust what moves within you at the deepest level. You are a vehicle for communication, after all, and giving voice to that which moves within you and seeks expression is your most important and most intimate work.

You are so much more than you had ever imagined, so much more important. So much more depends on you that you ever dared to consider.

As an emerging evolutionary catalyst, you need to be very clear about what's being asked of you, and where you are in the process of responding. This is not a chance encounter. It's destiny, and your response must be deep and authentic—and it can only be expressed and lived over time, perhaps over a greater span of time than you had ever imagined before.

In essence, you are the very evolutionary force that I have been speaking of throughout this book. This is who you are, just now beginning to unfold in your life. You are becoming the force that heals and joins, rather than merely one of its participants.

A *fear* often comes up here, that yielding to this force will lead to some kind of sacrificial lifestyle—or even the fear that you would gladly give up just about anything in order to live a life completely dedicated to a single purpose.

Sometimes this fear holds us back. We hesitate at the most crucial moment, for we sense that it means surrendering to something greater than ourselves. We fear that if we yield to the deep evolutionary impulse moving within us, we will lose ourselves. But actually, it is in responding to this deep calling that we find ourselves.

It's not so much that you become willing to take enormous risks in order for your life to matter. It's more that you realize that *everything* is at stake, and that together we could lose all that truly matters on this planet—so you are compelled to do everything you can to ensure that what is most important and most sacred about being human not only survives, but ultimately thrives. Accepting this responsibility and following what most deeply motivates you is the nature of being a r/evolutionary foodshed catalyst.

House on fire

Sometimes when I tell people about the coming global food crisis and the likelihood of losing fifty to sixty percent of food production capacity by 2050 because of global warming, I see that they just can't take it in—even people who are deeply involved in local food work. They seize up and tune out—not because they don't "believe" it or trust what I'm saying, but because they are existentially challenged. That is, they can't imagine there is anything they could *do* about this that might make any real difference.

But if someone tells you that your house is on fire, you will immediately spring into action—especially if you know that people or animals you love are inside. You will do anything, risk everything, to mitigate the loss.

This situation is no different. Our house is on fire. If we allow ourselves to know this, and let it into our being, then we will have no choice but to immediately spring into action.

As foodshed catalysts, we have no choice but to allow ourselves to know this. We cannot avoid this.

But what could "springing into action" actually look like? What would it look like to mobilize the local food movement in your community to a new level of radically increased impact, effectiveness, and scale? What would it take?

How can you radicalize your own contribution to the local food revolution?

How can you allow yourself to *feel* that our house is truly on fire, that you are desperately needed, and that you have something crucial to give in this situation? How do you come to accept that you are that important, that powerful, and that the situation really is this urgent?

What will happen to you when you truly step into this? What is the process you will go through? What happens to your relationships? What happens to your life?

And once all this sinks in, how do you keep from going crazy? How do you maintain your center? And how do you keep going when all seems futile?

Facing this and responding is the watershed moment in your life–your evolutionary turning point.

Whether your life in the end will truly matter depends on how you handle this now.

Being a revolutionary

Those of us following this path are revolutionaries.

Motivated by unseen forces, for reasons that none of us can explain, like those who initiated the American Revolution, we are the ones who are moved at a very deep level to "pledge our lives, our fortunes, and our sacred honor" to the most important and most urgent cause of our time: the localization of our food supply, the building of resilient food security and food sovereignty for us all–the reclamation of our future.

We've been recruited. This is what we've volunteered for. This is our destiny. We are among the initiators and early leaders of the local food revolution.

We are among those who truly know in the core of our being the sacredness of life and the sacredness of being human.

We rise up mysteriously from the exhausted soils of community, both rural and urban.

We are swiftly and unexpectedly transforming fractured landscapes, lovingly cultivating and rebuilding depleted soils, recovering and regenerating community in ways that kindle imaginations and infuse hearts with hope.

Together, we are planting seeds of aliveness that are contagious, seeds which are taking root everywhere in the almost forgotten yearnings of people's hearts.

Together, we are slowly birthing a restorative economy that is so radically different from what our society has become that to others at first it might seem utterly alien and impossibly unrealistic.

But underneath it all, in this process of making food local once again, in this meandering path *we are finding our way home*.

As this world plunges headlong into an age of Anthropocene darkness, we somehow know that out of the ashes of a dying civilization will emerge something new and unexpected. We know that it *will* arrive, for in our devotion and commitment we have preserved the very conditions conducive to life, and have prepared the soil in which these precious seeds can be planted and at last bear fruit.

Looking ahead

Some of us who understand that we are evolutionary catalysts find ourselves confronted with a wholly new challenge. Somewhere along the way, we shift away from what others may consider "front line" work to the realization that it falls to us to mobilize the revolutionary cadre of foodshed catalysts who are truly the only ones who can carry the work forward.

I have been going through my own initiation, and this book is a manifestation of that process. Co-creating the Local Food Academy is the next phase for me, an essential strategy in building our collective lifeboat through the Anthropocene. We are creating an opportunity to join a community of evolutionary foodshed catalysts in a project and journey on which the future of life on this planet depends, on which the evolution of life in this part of the universe depends.

Together, we are becoming evolutionary permaculturists, transforming our planet into a forest garden. We are terraforming Earth. This planet will become the generative center of aliveness that is its destiny.

This planet is our collective project. We join with others who realize this, and we commit to learning together, and sharing what we learn–making available what we learn as a kind of "seed bank" that will have value far beyond this planet or this eon of time.

#

Chapter 8

Preparation (Stage Five)

It's too late to solve global warming. There is no way to avoid the catastrophes of the Anthropocene, including the collapse of our current dominant civilization.

Our work now is to mobilize our communities to rise to the greatest occasion in human history, to create resilient regional foodsheds that will be our lifeboats through the dark times ahead and ultimately become the seeds of a new civilization.

This is a vast project, the equivalent of the kind of mobilization we saw in this country during World War II–but far more radical, far more revolutionary. The very future of life on this planet is at stake. The survival of the human species is at risk.

But what can ignite the local food movement into a local food revolution? How do we do this?

I'll take the risk of disappointing you by confessing that *no one has the answers.*

It's true that no one has ever localized a regional food system before, certainly not at the levels of population that we are faced with these days. No one currently knows how to do this.

But with an understanding of how emergence works–and how healing, restoration, and regeneration unfold–we can be confident that embedded on this planet and in our culture are individuals who carry the seeds of the local food revolution within themselves, whether they recognize this or not. You are among them, which is why you are reading this book.

You are called to lead the local food revolution in your community.

What's happening with you here is radical. You, like other emerging foodshed catalysts, are being transformed, activated, radicalized, and you must allow this to happen. The revolution begins within.

The American Revolution

In studying the history of revolutions, I've been deeply struck by Chris Hedges' provocative book, *Wages of Rebellion*, especially where he says:

"There is nothing rational about revolution. In the face of insurmountable odds, revolution is an act of faith, without which the revolutionary is doomed. This faith is intrinsic to the revolutionary the way caution and prudence are intrinsic to those who seek to fit into existing power structures. The revolutionary, possessed by inner demons and angels, is driven by a vision. I do not know if the new revolutionary wave and the revolutionaries produced by it will succeed. But I do know that without these revolutionaries, we are doomed."

Were it not for such revolutionaries who have gone before us, none of us would be here today.

This nation was born as the result of a revolution. Revolution is our heritage.

In the last few years, I've had the great privilege of going back to study the American Revolution as an emergent process. I've come to learn much about the individual who ignited that revolution, Thomas Paine, whose essay "Common Sense," published in January 1776, galvanized the people of the 13 colonies to the point that they took the historic and very dangerous step of declaring their independence from England and establishing a new democratic nation.

The spirit that inspired people in early 1776 is mysteriously alive today, and we can feel it emerging in the local food movement.

Revolutionary failure

The American Revolution was evolution or emergence at work, bending towards justice. But in many ways, that revolution failed. It gave birth to a deeply flawed democracy rooted in greed and exploitation, built on the backs of millions of slaves abducted from Africa, built on the utter decimation of the indigenous peoples of this continent who had lived here for millennia before us. Those gaping wounds have led to the distortion and ultimate death of democracy in this land, and the utter corruption of its resulting economy.

The British Empire may have lost the American colonies, but the new nation was eventually colonized by the temptations of empire itself.

In a certain sense, we can say that all revolutions in human history have failed.

I was astonished to learn that Thomas Paine, who was raised in a Quaker family, was an abolitionist. Part of his vision for the new nation was the elimination of slavery. How painful it must have been for him when that calling could not be fulfilled in his lifetime.

Paine was later ostracized and essentially banished from the new nation for his stand on this issue. When he died in New York in 1809, he was penniless and an outcast, nearly friendless. Only six people came to his funeral, and three of them were black.

The official end of slavery, to which Paine had dedicated his life, did not come for nearly a hundred years, in 1865, with the passage of the 13th Amendment, under President Abraham Lincoln. And while slavery was abolished legally, we know that it still exists here and around the world in many different forms.

That evolutionary impulse, along with many others emerging some 240 years ago, went largely unfulfilled with the American Revolution and with the nation that it birthed.

Today, at the heart of the local food movement we find an enormous impulse for justice, for sovereignty, for freedom. But this impulse has not quite found its voice, and has not yet unified the movement.

Through the advent of the industrial food system, justice, sovereignty, freedom, and democracy have all been usurped and stripped away from us–stolen.

I have long been disturbed that the proliferation of industrial agriculture in the developing world was called the "Green Revolution." This was not a revolution at all, but a coup. Historians in the future will say that this was the moment we decisively lost control of our food supply.

It is through the local food revolution–the radicalization of the local food movement–that we will begin to restore justice, sovereignty, freedom, and democracy in this land. It is through the local food revolution that we will reclaim our future.

Getting started

At this point, you may feel compelled to begin hatching and organizing projects to mobilize your community around this effort. But the reality is that you are not quite ready to move into action.

We all want to know what we can do in our community. Where do we start? What project can we launch? How do we do this? What solutions are available?

But what if the "solution" is *you* and what you yourself are ultimately able to catalyze?

Healing, restoration, and regeneration are of course the long-term solution. But they require *your* taking an active and conscious role.

As a foodshed catalyst, what you do has the potential of ultimately impacting millions upon millions of acres of farm and ranch land around the world, and millions of people. This is the greatest transition since the domestication of plants.

The revolution must first begin within yourself. *You* must become radicalized, internally revolutionized. This is a transformational process, and it takes some time. You cannot make this happen.

Only when you have gone through this process and have come to the place where you can dedicate your life and your resources to this cause will you be able to take up the question of *how* to ignite the revolution in your community and mobilize the great effort needed to build a network of highly-localized regional foodsheds.

Until you have yielded your very life in service to this, setting aside your own preferences and ambitions, you can only add to the chaos and confusion and separation and conflict that so characterize our evolutionary bottleneck.

You must come to the place where you choose to yield to the evolutionary force that is moving within you. *You choose this because you have no other choice.*

The calling comes when the world's great need and our deep inner longing finally intersect.

Beginnings

Now, if you know that localizing your food supply is going to be essential for your community–your local population, your region, however you define it–what do you actually *do*?

While it's true that every community is unique, there are some things about the process that are going to be very similar anywhere we try to localize food.

We're discovering that food localization is a natural evolutionary process of healing and regeneration and co-creation–and that it's learnable, replicable, flexible, adaptable, and systemic–and hopefully even contagious.

The process of localizing a regional food supply unfolds in identifiable, cumulative stages. Fundamentally, the process charts the shift from a region's reliance on the global industrial food system to the establishment (and even dominance) of a fully-integrated regional foodshed.

This is an emergent process, and *cannot be made to happen*–any more than farmers can make plants grow and bear fruit. This is crucial to understand. We cannot make food localization happen, and we can't control it. Any efforts to do so will end in failure. All we can do is catalyze and nurture the process. And that can make all the difference.

This will likely be a process spanning generations. We are among the initiators and early adopters of this process. We are evolutionary catalysts–specifically, foodshed catalysts.

Early stages of the process

Within the context of this book, I can only share a broad overview of how a highly-localized foodshed emerges in a region, arising out of the industrial food system that has dominated the food supply for decades.

It's a process, which unfolds in seven distinct stages. The entire process is occurring in the context of a dying civilization and a world that is plunging into the Anthropocene epoch of consequences.

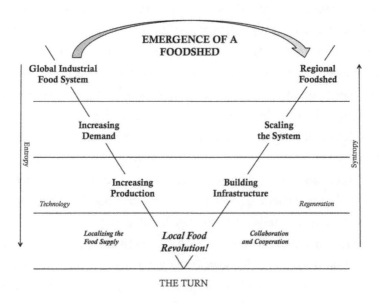

Stage one represents the status quo, dominated by the global industrial food system. Here, the region's foodshed effectively stretches around the globe, with almost none of the food that people there consume being produced locally. Nearly all the farmers and ranchers in the area are devoted to producing commodity crops and livestock, almost all of it for export.

In *stage two*, demand for local food begins to grow. What begins to emerge here is a community of eaters, people who are enlivened by the experience of local food and all that it represents.

In *stage three*, production begins to grow to meet growing demand. Producers and eaters and commercial food buyers alike begin to discover that the hunger for wholesome local food far exceeds supply. Soon everyone discovers there is simply not enough non-industrial local food available to meet the growing demand.

In *stage four*, the turning point in the entire process, obstacles and challenges abound. Those involved in the emerging regional foodshed begin to recognize that the local food movement cannot viably evolve in the region without radical, systemic change on a much bigger scale. This often seems a daunting prospect. This stage will either result in a local food revolution, or the complete success of the unholy alliance. Many emerging foodsheds in the U.S. are deep in the throes of stage four. Seeing this helps to make sense out of what is happening.

During *stage five* it becomes essential to consciously develop the local food supply chain, the heart of the regional foodshed. The focus in this stage is to catalyze an integrated, collaborative, values-based system to connect producers and food entrepreneurs with their larger commercial markets. This means rebuilding the missing local food infrastructure of aggregation, storage, processing, distribution, and marketing—in the form of a network of local food-related enterprises.

Stage six of localizing a foodshed is to significantly expand the regional food system, to ensure the long-term resilience of the foodshed. If a local food system is to feed a significant portion of its population, reaching levels of 25 percent food localization or more, it will be necessary to achieve a level of scale and efficiency far beyond what we're seeing almost anywhere in the U.S. today. This stage is unknown territory, where nobody in the local food movement has gone before, and it may be decades away.

Stage seven is where evolution is ultimately taking the process, where we will have catalyzed a regional foodshed that is economically robust, environmentally sustainable, resilient, and self-reliant, that ensures food security and food sovereignty and food justice for all its people, that contributes to the health and happiness of our citizens, that revitalizes our local economies, and that regenerates our bioregion. The foodshed that finally emerges is a living being, regenerative, itself capable of proliferating other similar living beings. This is what the local food revolution is all about. This collective effort, taking place in regions across the continent and around the world, ultimately ushers in the next phase of human evolution, the Ecozoic Era.

This process isn't going to be easy. The challenges are formidable, and the task of taking back our food sovereignty and reclaiming our future is a very long-term project which will take generations, perhaps millennia, much longer than any of us realized when we began more than a decade ago.

But the local food revolution, ignited by foodshed catalysts everywhere, is now planting the seeds for this radical shift in how humanity feeds itself.

Out of the ashes of a dying civilization something entirely unexpected can arise.

None of us here will live to see this new civilization, but we can be part of the effort to plant and cultivate its seeds.

In *Tomorrow's Child*, Rubem Alves writes:

"Let us plant dates, even though those who plant them will never eat them…. We must live by the love of what we will never see. This is the secret discipline… a stubborn commitment to the future of our grandchildren. Such disciplined love is what has given prophets, revolutionaries, and saints the courage to die for the future they envisaged. They make their own bodies the seed of their highest hope."

This is the spirit of the local food revolution.

Developing a plan

I've very briefly described the naturally occurring process of food localization for two reasons. The first is to give you a sense of where your community is now in the process (most are attempting to increase local food production in stage three, or have entered into the chaotic and confusing challenges of stage four). The second is to give you the clear sense that finally getting to a highly localized regional foodshed is a very long process but cannot be forced.

At this point, you do not yet know what your specific role or mission in this process will become. For now, you have already accepted that you are called to lead the local food revolution in your community, to be a revolutionary foodshed catalyst. But you are just now embarking on the journey of discovering what this actually means for you.

Meanwhile, even though you know very well that the situation is urgent, you must first *prepare* yourself for what will be a very long journey.

Your first responsibility as an evolutionary foodshed catalyst is to develop a plan–not for the localization of the food supply where you live, but to prepare for your own development as a revolutionary leader in the process. You must prepare yourself physically, mentally, relationally, and spiritually. You must take on the discipline of becoming an evolutionary catalyst, which for you will be a lifelong process.

Your plan will significantly change over time. This is natural. Your ability to see and feel will mature as you proceed, and this will affect your plan. Conditions will change, from the local to the global, and you must adapt accordingly.

Here are some recommendations to consider as you begin your planning process:

- Cultivate your ability to see how healing and regeneration are beginning to manifest in your community, and how you might be able to quietly support these trends. Adopting a spiritual practice can be useful in this regard.
- Find and connect with centers of aliveness emerging in your region. Become involved where appropriate, contribute where you can. Allow your search for such contagious centers and your learning to go beyond your home region.
- Develop relationships with a variety of farmers and ranchers in your area, along with food entrepreneurs and food service providers, and seek to become a part of their world. Encourage and support the emergence of agrarian culture.
- Look for other emerging or potential foodshed catalysts in the area where you live. Invite them to join you. Consider catalyzing a core group of similarly oriented people to explore how you might be of service together. This can become a local *community of practice.*
- Take very good care of your physical body (you'll need it for a long time), with exercise, sunshine, and plenty of healthy local food. Eliminate industrial foods from your diet as much as possible.
- Constantly deepen your understanding of the evolutionary process, so that you will increasingly be able to apply it in your own life as well as in your emerging foodshed.
- Attune to what science is learning about the unfolding of the Anthropocene Epoch, now in its very early stages. Pay

close attention to trends in global warming. This will require disciplined study.

- Investigate the network of relationships that support the industrial food system in your area, and learn what you can.

- Develop whatever skills and resources you will need to ensure that you will be personally resilient and self-reliant in the coming decades. Align your income-generation so that it supports your development as a foodshed catalyst and does not detract from this direction.

- Consider ways to financially support and even capitalize the food localization effort in your community.

- Be very cautious about joining or initiating specific food-related projects. While these may provide useful learning experiences, you must take care that they do not become distractions from your larger purpose–which is to prepare yourself for leading the local food revolution in your community over the long term. At this point, you do not even know what evolutionary leadership really means, and you have many capacities to develop before you can meaningfully contribute. This will require great patience and constraint on your part, and compassion for yourself as well as for others who are attempting to support local food in your area.

- Do not doubt that the region where you live now is where you are most needed, that this is your place of service. Do not become persuaded that another region would be better for you.

- Seek guidance in evolving and implementing your plan over time, but remember that your deepest feelings and truest desires are your most reliable guides.

- Refrain from seeking recognition or acknowledgment for the work you might do. You serve the evolutionary process itself.

- Avoid the temptation to become a local food activist. This is but another form of attempting to make things happen. You are an emerging foodshed catalyst, which is a completely different way of working.
- Create a personal notebook where you will write down the various iterations of your plan and track daily what you observe and learn along the way.

Be kind to yourself. Allow yourself sufficient time and a supportive environment to do this planning work, over a matter of months. Do not try to rush this part of the process. Your developmental plan will be completely unique, and you are not obligated to share it with anyone else. At least in the early stages, it is not necessary for anyone else in your community to know that you are taking this journey.

Chapter 9

Implementation (Stage Six)

I n this chapter, we will explore the process of your own emergence as a leader of the local food revolution in your community.

I'm talking about a very different kind of "leadership" than what you might expect. Here I'm speaking of leading by essentially being an evolutionary catalyst.

This means that you accept responsibility for localizing the food supply, recognizing that whether or not this ultimately happens in your community really comes down to you.

This does not mean that you will do this alone. Nor does it mean that you will ever say to anyone that you have accepted responsibility for food localization in your area, nor that you will ever be recognized for your contribution.

It is as if you are the sole outpost of the evolutionary process where you live. You are the one who clearly sees the urgent need for localizing the food supply. You understand the nature of the human predicament on this planet, and how food is the pivotal issue.

You know in your heart that a highly localized foodshed is the only pathway to a viable future for the people in your region.

As you begin, you may be the only one who sees and feels what you see and feel, who knows and understands what you know and understand.

It does not matter if you have been involved in local food work before. It does not matter how many others in your area are working in the arena of local food, or how long they have been engaged. You are beginning something entirely new.

———

Years ago, I learned about an international religious organization that trained its evangelists to be prepared to be deposited on the outskirts of a town or city, where they were given a bicycle, $50 in cash, and one suitcase or backpack for clothing and personal necessities. The evangelist was expected to enter that community, ethically develop a means of financial support, begin teaching, and ultimately develop a teaching center that could accommodate many people, ideally purchasing the property in the name of the organization. It was astonishing to me how often these evangelists were successful, starting from scratch, and helped me to see the extent to which we underestimate our capacities.

As a foodshed catalyst, perhaps you will not be starting so completely from scratch, but it will probably feel like starting over.

You are the seed of the healing, restoration, and regeneration that will unfold in your region through the process of localizing the food supply. You are an evolutionary catalyst.

This is very different from being an activist or change-maker. Here you are allowing yourself to become increasingly aligned with the evolutionary force itself. This is what you serve, and it moves in you and through you.

It may be some time before anyone knows that you are working on this level. Your work will likely be invisible at first.

Holding the vision

You alone hold the vision for the regional foodshed that is needed. This is where you begin. The vision has come to you. Others may see aspects of the vision, but only you have this unique evolutionary perspective. You cultivate and nurture this vision as best you can, realizing that your initial grasp of the vision will grow and evolve over time. At first, you are seeing the broad outlines of the vision, not the specifics. But the specifics will gradually become clear.

You are the vision keeper for the potential of food localization where you live. This is something like being given the tribe's sacred bundle, which must be protected and preserved for generations to come. Being the vision keeper is a sacred duty.

Cultivating a core group

However, the vision cannot manifest if you carry this sacred treasure alone. You must share it with others.

And there are others in your community who are called to this work, just as you have been. It is your job to find them and to communicate this vision to them in a way that activates them deeply. This may take a long time, but eventually you will be able to assemble a small group of people who are able to hold the same vision that you do, and together are willing to accept responsibility for its fulfillment.

In everything you do, you must constantly be on the lookout for these people. The process of food localization will simply not be able to unfold in your community without them. You will be taking this journey with them for decades to come. You are entering into an adventure in collaborative co-creation.

Assembling this core group of foodshed catalysts is a process that cannot be rushed, and must be done with great care and discernment. It may take years for you to find this group and discover the ways that it can best function.

And here I will say something that may be disturbing to you, that it is entirely up to you alone who will be invited into this core group. And it is entirely up to you to determine if someone should be excluded or dismissed. This is a great responsibility, but–at least in the early years–it must be yours alone. You are the one who has been entrusted with the sacred vision and are responsible for its fulfillment. *This is not about outcomes, however.*

During this stage, you will learn a great deal about communicating the vision that you hold. This is vastly different from sharing information. This is communication which is transformative. This is *transmission*, allowing yourself to be a vehicle for expression on behalf of the evolutionary force itself.

The maximum size for this core group is probably eight people, including yourself. But once there are three of you in this nascent group, you can begin working together. You may invite others to join the group after you have begun.

Understanding current reality

Your first project as a group is to come to a shared understanding of current reality.

Here the overarching task is to discover everything you need to know to successfully be catalysts for localizing the food supply in your region–together.

You begin by sharing with the group everything of relevance that you have discovered. You will share what you have learned about the nature of the human predicament, how the industrial food system is the primary cause, and just how important and urgent it is to build regional foodsheds.

But you will also share what you are learning about the process of evolution itself, how healing and regeneration unfold and how as individuals (and as small groups) you can align with this process and become its vehicle for expression.

Fortunately, you do not need to be an expert on food localization or evolutionary process. Remember that no one has ever localized a region's food supply before, at least not with the population levels that we face today. There are no "experts" in food localization, only pioneers who are *learning* to do this, carrying a vision that was imparted to them.

Sharing with the group what you know and understand will initiate a process of exploration and discovery as a group.

This process is first a matter of seeing, feeling, and understanding *together*. For this to happen, you must work together to answer many, many questions until you arrive at a shared sense of current reality–from the big picture of the world today, to a candid assessment of the conditions in your own region.

Again, the overarching question that you're addressing is this: *What do we need to know, and what skills and resources do we need to develop, in order to bring this vision of our regional foodshed into reality?*

How deep this process of exploration and discovery goes and how long it takes will depend on the nature of the group itself and the people within it. It's also essential to remember that current reality is constantly changing.

Along the way, you will discover that for individuals to participate in this process successfully, they must be able to abandon beliefs in favor of acknowledging reality. They must be sufficiently healed and whole in themselves so as to not be obstacles to what is attempting to unfold within the group. They must be committed to learning and knowing and seeing and feeling together. They must be willing to yield to the evolutionary process itself.

It has long been my sense (my own vision) that such a group could ultimately be entrusted with the vision and responsibility of the healing and regeneration of life on this planet. Your group may ultimately be called to that level, but you begin by focusing on the vision of your regional foodshed.

—

You need to be aware that the most important outcome of this part of the process is that the group is becoming a living organism. In a sense, you are developing a shared identity–shared vision, shared purpose and passion, and shared reality. These are the foundations of true unity.

Your group is the building block of the kind of community that we all long for.

The Turning Point: decision and commitment

There will come a point–and you will feel it together–when the pursuit of current reality will reach its zenith and its overarching question will become: *What must we do?*

To state this emerging question a little more explicitly: Based on what we know is needed, based on what we understand of current reality, based on who we are together, what must we now do?

Here the process begins to move into the realm of action, work that can directly contribute to the localization of the food supply.

Getting to this point can take quite a long time, and it's important to resist the temptation to begin developing plans and projects too early. Actions that are not grounded in the vision and rooted in current reality, and supported by co-creative collaborative relationships, will lead to failure.

At this stage, the process becomes focused on making decisions and commitments.

The first decision for the group to make is whether together you will take responsibility for cultivating the vision and being a vehicle for its manifestation into reality. *Will we do this?*

Not much of value can happen until you confront this question together. But once you have made this life-changing decision together, the evolutionary force can begin to move powerfully within the group. This is a momentous occasion, and marks nothing less than an evolutionary breakthrough on the planet.

What you're learning here is that it's impossible to be an evolutionary foodshed catalyst alone. Only when the seed of the vision is planted and nurtured by a small group of people together can the process actually happen.

—

You'll likely be able to see signs of this evolutionary force attempting to emerge in many other projects and individuals. But if they are not joined in the kind of co-creative collaboration that your group has, guided by vision, current reality, and an understanding of the process by which evolution/healing unfolds, then the seed is not taking root with them.

On the other hand, if it appears that they might truly be the very kind of group that you are attempting to catalyze, then you should consider joining them. If you do, you'll soon know if the fit is real.

—

With the decision to take on the responsibility of manifesting the vision of a regional foodshed, your core group is embarking on a journey that will be central to each of the individuals involved for the rest of your lives. This is not a commitment that can easily be rescinded. This is a commitment of a lifetime, and perhaps beyond.

So, again, do not rush this part of the process.

And before you move on to consider specific projects and actions, you should first develop a written declaration that expresses precisely the decisions and commitments that this group is making together.

Once this is done, the focus of the group will shift to a consideration of what specifically the group will decide to do together–what action to take.

Here I would recommend that you not take on the biggest or most urgent project in your emerging foodshed. You are initiating *a learning process* here, and your group will need to practice together the role of being a catalyst for the emergence of a regional foodshed.

Start small and experiment.

Circle of discovery

The group must decide together where to begin, what to tackle. For instance, you might choose to help significantly increase sales at farmers markets. Or initiate a community-wide "victory garden" project. Or build an investment fund to help capitalize local food and farming enterprises. You will need to choose a starting place from among a multitude of possibilities.

You'll know where to begin because a vision for a project or action will naturally emerge within the group, and it will be communicated so powerfully that the entire group will be able to feel that this is the appropriate direction for the time being.

With this project vision, you will be able to discern its estimated minimum design requirements–the people, the resources, and the time that will be needed.

Once these things are clear enough that the group feels ready to proceed, it's time to launch the project or activity.

With the vision and parameters for the project in hand, your group can begin moving toward bringing this specific vision into visible form, moving forward as a body.

For the project to flow smoothly, you must give careful attention to the relationships involved–with the members of your group, and with members of your community who may become engaged in the project. Effective relationships depend on alignment–with the vision, with the plan as it has been developed, with those engaged in the project together, and with the intended result.

No matter how well planned a project may be, no matter how brilliantly conceived and designed, things will go awry. Estimates and projections will prove to have been inaccurate. Assumptions will be challenged and changed. Accidents will occur. People will make mistakes. And the project plan will have to be adjusted, perhaps even abandoned.

Giving birth to a new future is, after all, an uncertain and unpredictable process of emergence. It's crucial to pay attention to your group's experience and feedback during the project and to learn from it.

Your core group will communicate frequently during the course of the project to assess progress and make necessary course-corrections.

At the conclusion of the project, your group will reconvene to *celebrate*–to reflect on what has been learned and experienced, to examine what has been accomplished, and to consider together what needs to happen next.

This brings this part of the process to full circle, putting your group in a position to move to the next step.

Following this circle of discovery is fundamentally how your group will function together. It is in being engaged in your community in this way that you will discover together how to be vehicles for the emergence of your regional foodshed.

Sustaining the group

This is very challenging work, and will ultimately take everything you have to give. But if this is what you're called to do together, nothing else will satisfy you like this work.

As the initiator of this group of co-creative collaborators, you bear primary responsibility for the health and well-being of this emerging organism.

Here are some recommendations:

- Keep the learning going. Initiate projects that you would like to directly participate in yourselves. Support each other in taking relevant classes, courses, and workshops. Share skills, books, articles, and important insights.
- Develop activities that deepen you and the group. Consider exploring Joanna Macy's *The Work that Reconnects*. Enjoy renewing activities together like outdoor experiences, and shared meals with no agenda (especially potlucks).
- Review your balance of giving and receiving. Take steps to keep yourselves sustainable, resilient, and self-reliant.
- Deepen your own ongoing preparation (inner and outer).
- Remember, the work you are doing is on all levels–body, mind, soul, and spirit.

Your initiating group will evolve over time. Inevitably, people will come and go. If people leave, grieve and move on. If you invite new people to join you, take great care to integrate them into the group.

Becoming an authentic voice

This work may or may not need you to be an effective public speaker. But if you don't take on that role yourself, someone else in your group will need to become your spokesperson.

However, it is necessary for you to become an effective communicator, though perhaps not in the usual sense. You are becoming a vehicle for *transmission*, allowing the evolutionary force to move in you and through you. This, I suspect, is the deeper meaning of developing an "authentic voice." This is what will make it possible for you to learn to be an evolutionary foodshed catalyst.

Being a leader

Being a leader in an evolutionary sense–an evolutionary catalyst– is a spiritual discipline that you will learn to cultivate and practice. You will learn to become attuned to the evolutionary force that is flowing in you and in your community.

You have the distinct advantage of being able to see and understand that food localization is a process with discernible stages. With this framework, you can discern where the process has gotten to in your community, where it's going, where it's stuck, and what's needed. You will begin to learn the complex patterns of relationships that emerge at each stage, and how you can be of service to these relationships. This perspective puts you in a unique position in your community, and as you share what you're learning with the other members of your core group, they will come to occupy this same kind of position.

It's not necessary for you to know all the facts about everything. No one is an expert in food localization. But knowing that this is a process of healing and regeneration puts you far ahead of nearly everyone else. Your challenge will be to effectively communicate what you see and what you know.

You become a true leader by being engaged on the front lines of the struggle for freedom and sovereignty. You learn to lead from the front and from the rear, and even from the middle.

You lead by holding the vision, cultivating relationships, staying attuned to current reality, and making decisions not out of fear but out of service.

Deepening your learning

It will serve you in your work of food localization to learn the principles and practices of permaculture, and to become engaged with the growing global community of permaculture practitioners and teachers. While this will give you deep and practical insights into how food can be most appropriately produced in your region, it will also give you a profound education in how life actually works on this planet.

As you learn permaculture, you will learn about "pattern language," which was pioneered by Christopher Alexander. Exploring pattern language is extremely valuable in the realm of permaculture and human relationships. But keep in mind that Alexander's classic work in pattern language was published some 40 years ago.

His magnum opus, going far beyond pattern language, is *The Nature of Order*, an unprecedented four-volume exploration of how we as individuals and groups can align with the underlying evolutionary process. I recommend that you and your core group take time to study this magnificent work together. It has the power to guide you in profound ways into the discipline and practice of being an evolutionary catalyst. But fair warning, allow years for this study. Take it in bite-sized chunks.

Become a student of evolution, and a student of the just-emerging Anthropocene Epoch. You will need to bear witness to the unfolding of both these aspects of the human experience in the coming decades.

You will likely be the single person in your community who has most fully integrated these frameworks, both crucial to the long-term success of your regional foodshed.

And always be a student of food localization. We're all learning here.

It will inevitably fall to you to speak of difficult things–difficult mostly because they are so devoid of beliefs and cultural accretions. You will become a carrier of wisdom, and you need never be afraid of fear.

Always remember, you are doing this work out of love, out of a devotion to service.

How you live, how you relate with your fellow humans, and how you relate with the larger community of life are far more important than anything that you might accomplish in this lifetime.

Your preparation will continue long after you leave this life.

You will never not be a student.

You will never not be an evolutionary catalyst.

You are just beginning, discovering how to be a vehicle for food localization in the place where you now live.

Guidance

Seek guidance, not from experts but from the wisest elders of our human tribe, those who have deeper connections to what is sacred, deeper connections to life itself, those who can see beyond the urgencies of the moment to behold the broad sweep of evolution itself. Seek their moral and ethical and evolutionary and spiritual perspective.

Find and connect with those who have gone before you, and ask for their assistance. They are waiting for you. Honor them by integrating their guidance into your learning and practice.

Develop the patience and compassion of saints and angels. The Anthropocene will be a very difficult time for the human species (and for all life on this planet), and you bear great responsibility.

#

Chapter 10

Realization (Stage Seven)

You and your core group will never reach the point where your work is complete. You will never see the day when the food supply in your community is fully localized, for this will take generations to accomplish. You are merely initiating the process, and serving as catalysts for food localization. In fact, it will be very useful to consider yourselves an *initiating group*, and yourself as the initiator.

Your function is not to organize the food localization effort in your region, but to serve and support the process itself, and to serve and support those in whom the process is coming to life. This is a very different approach to leadership.

As an initiating group, you are calling upon farmers and ranchers and food entrepreneurs new and old, large and small, to join the local food revolution. You are asking them to see themselves as social entrepreneurs, to become drivers of systemic change, to seize the opportunities that food localization presents, to innovate the new systems and processes that will be needed, and to forge high standards for local food production, as well as high systemic standards of fairness, equity, sustainability and stewardship, ethics and accountability, cooperation and collaboration, and food sovereignty.

You are calling upon chefs, restaurateurs, caterers, and food retailers–who are in a position to influence more eaters than almost anyone–to source a substantial percentage of their foodstuffs from local growers and value-added producers.

You are calling upon conventional food producers, manufacturers, and service providers to diversify their operations by beginning to meet some of the growing demand for local food.

You are calling upon eaters to participate in this historic revolution, which will help turn our troubled world right-side up. You are asking them to shift their eating, cooking, and food purchasing patterns, to become homegrown food producers, and to shift some of their money into the local food economy. You are asking them to abandon the story that they are consumers and to begin taking up their roles as engaged food citizens.

You are calling upon local citizens to activate their economic power and to move a small portion of their money into the emerging local food system.

You are calling upon institutional food buyers–schools, corporate cafeterias, airports, convention centers, etc.–to commit to devoting a significant portion of their purchasing budget to local food.

You are calling upon local government leaders, policy makers, community activists, and nonprofit agencies to raise awareness, to mobilize public support, to focus priorities, and to help remove barriers.

You are calling upon holders of significant financial wealth–individual investors, successful entrepreneurs, philanthropic foundations, capital-investment firms, hedge funds, banks, credit unions, and large corporations–to devote a portion of their capital to helping build the regional food system.

You are calling upon experienced legal counsel, entrepreneurs, business and agricultural consultants, and experts of all disciplines to step into the fray to help–without demanding traditional pay rates.

You are calling for journalists, creative writers, media producers, bloggers, artists, and musicians to investigate, capture, and celebrate the local food revolution, bringing it to center stage where it belongs.

You are calling upon faith communities to join together in supporting the local food revolution.

You are calling for a rebirth of food-based culture and community.

—

As a foodshed catalyst, a key responsibility that you must also accept is to find and recruit and support others who are being called to this work, in your community and beyond. You are recruiting people into aligning with the evolutionary process that is manifesting healing and regeneration on this planet. This is something like inspiring growers to become seed savers.

Food as a gateway

When Lynette Marie and I began to explore initiating the process of relocalization in Boulder County, we viewed the work as a multi-pronged effort that would separately tackle such issues as food, economy, energy, transportation, emergency preparedness, health care, and culture. But we soon came to realize that food is *the gateway issue*. Without the foundation of a localized supply, a community can never achieve resilience or self-reliance. We came to see that localizing the food supply was the pathway to healing and regeneration in all aspects of society.

As a revolutionary foodshed catalyst, you are part of a much larger process that is unfolding on this planet. And while local food work will require much of you, it is possible that as you progress you will find that you are being called internally to something greater.

Serving as a foodshed catalyst is actually your initiation as an evolutionary catalyst. Your path may ultimately lead beyond the issue of food localization, but will always include food localization. This need not concern you now, but please remain open and attentive to what is wanting to emerge within you.

Over time, you will find yourself having a significant but often mysterious and invisible impact on the process of food localization in your community. You may never be recognized in your community as a leader of this effort, but you will know what you gave and what you helped catalyze with your initiating group.

As an evolutionary catalyst, you are not responsible for outcomes. The outcomes that are most deeply needed in a particular situation may not be what they appear to be. For instance, the success of your initiating group cannot be measured by the percentage of food localization achieved in your region over a period of time. Your real success can only be evaluated in terms of the capacities you have developed to align with and express the evolutionary force of healing and regeneration, and what you have learned about working together as a united body of co-creators. What is most important about going through this process is not what you accomplish, but what you become.

In a sense, the work of food localization is a training ground for the evolutionary catalyst. You are developing capacities and relationships that may ultimately be needed elsewhere. Trust where the process is taking you, and trust your ability to respond appropriately.

Legacy

As Lynette Marie and I can testify, in a much shorter time than may seem possible now, you will become an elder in your community.

People will gradually seek to learn from you, to draw from your wisdom and experience. It's important for you to prepare for this.

What you are learning now and in the decades to come from working to localize the food supply can have great value for those who will inherit this mission in the future. You are building a legacy that must be passed on. This will require conscious effort on your part.

Many farmers keep detailed notes on what happens on their farm, recording everything that they or their successors may need to know in the future. They are constantly learning about soil, weather, biota, plant and animal species, human interactions, and the entire web of relationships that makes farming both necessary and possible. They constantly refer to these notes from the past to help them understand what is happening currently, and to plan for the future.

You would be wise to follow this example, individually and as a group. Meticulously record your observations, your insights, your discoveries. You will need these notes to guide your way forward, and these notes will be part of your legacy in the future.

At this pivot point in human history, what you learn and experience will be of great value to those who follow you. You are evolution's arrow, and the trail of learning you leave behind will be of inestimable value to future generations.

—

What you are attempting may seem utterly impossible. But such is the realm of the evolutionary catalyst.

In an earlier global crisis, Winston Churchill is rumored to have said, "Leadership is the ability to go from failure to failure without losing enthusiasm."

This is the kind of leadership you will come to embody.

As foodshed catalysts, we may fail a thousand times before we ever experience a systemic success. And yet this is precisely how evolution works, relentlessly seeking to emerge through countless transitional entities. We ourselves are transitional entities, and we must be willing to be used for this purpose. We serve something greater than what can be measured or even visible in this lifetime.

You are part of the flow of evolution itself. And evolution is an unstoppable force.

Learning to see food localization as an emergent process of healing and regeneration liberates you from the crushing enormity of what you face. In seeing the underlying structure of this localization process–which mirrors emergent processes in nature–you can learn to identify where you are regionally, and begin to chart pathways through the complexities of the work, anticipating what will be needed in the future. You learn there is a kind of seasonality to all this, just as there is an appropriate time for soil building, for planting, for cultivating, for harvesting, and for celebrating.

You are learning how humans can collaboratively shape systems and structures which generate wholeness and aliveness. Your learning ground is applying these principles to the emergence of a localized foodshed, but you may be called to apply these principles in other arenas of human endeavor as well. Be open to this, but discerning.

You understand that the awakening of a foodshed is an event of great historical and planetary significance. You discover that food localization is the evolution of the universe itself at work in us and through us. In these troubled times, there is no arena of life where the potential for human evolution–and restoration and regeneration–is more timely or holds more meaning and potential.

Reclaiming the future

What is ultimately at stake is human freedom and sovereignty. Once you understand this, you have no choice but to do everything you can to reclaim the future. The destruction of the biosphere and degradation of our food supply must stop, no matter how long it takes, no matter the cost. You know full well this effort can only be realized over a very long period of time. You are initiating the mobilization that represents humanity rising to the most demanding occasion in history.

You are shaping the future of your foodshed, and it is shaping you. The rebuilding and localization of a regional foodshed is a massive process. No one can control this process, but you have the great privilege of being its catalyst and an ardent advocate for co-creative collaboration.

No one can say whether this effort will ultimately be successful in your region. No one can say whether the emergence of a network of regional foodsheds will occur in time to avert utter catastrophe.

But the evolutionary arc of the moral universe is relentless. Everything you contribute to this process is of great and lasting value.

You have the unique blessing and responsibility of being able to participate in earth's redemption.

You are initiating the reversal of the destruction of the last 10,000 years of industrial civilization. You are a vehicle for healing and regeneration. You are a part of a vast coordination of lives toward a single aim, the ending of separation and the inauguration of the new civilization that will be the fulfillment of the dream of the earth.

And here I can only offer my humble prayer in the hope that you will adopt it as your own:

May we align ourselves with that long evolutionary arc of the moral, sacred universe and end the reign of those who would control our food supply, those who would control the very forces of nature, those who would control what is most precious and sacred in our seeds, soils, and souls.

May we restore soils and souls and hearts and minds, and may we reverse this dreadful course that threatens to dominate our humanity and undermine our freedom.

May we truly be people of the earth, connected with land and water and sky and the natural cycles of life, connected with the seasons, connected with each other and connected with the greater community of life, connected with the sun and the moon and the planets and the stars and the galaxies, and may we be connected with that sacred evolutionary spark that dwells within each of us.

Together, we are the local food revolution.

#

Chapter 11

Challenges and Obstacles

A s you proceed on the path of being a revolutionary foodshed catalyst, working in co-creative collaboration with others who share your vision and commitment, you will inevitably face difficult challenges and seemingly insurmountable obstacles. Some of these are predictable, but some may go beyond current human experience.

As no one has successfully localized a regional foodshed before, you will have little previous experience to draw from. You will need to rely on your own inner compass as much as possible, and seek assistance when necessary.

But in this brief chapter I will address a few of the challenges and obstacles you are likely to face.

Who am I to do this?

For reasons that are beyond your comprehension, the universe has chosen you to pioneer the localization of the food supply in your community, to catalyze the emergence of a regional foodshed. Your acceptance of this role will at first be somewhat tentative, and at times you will come to doubt whether you are the right person for the job.

Self-doubt is normal, but you must learn that ultimately you have no choice in the matter. You must accept that this has come to you and that you must do the best you can. You cannot walk away without causing great harm to yourself and to the flow of evolution itself. Much depends on you, far more than you can now imagine.

It is useful to remember that you will never be in a position to be able to accurately evaluate your progress or your emerging capacities. Only those who have gone farther on this evolutionary path can offer you useful guidance and perspective.

Premature and inappropriate commitments

You will feel inclined to commit to people and projects far earlier than is appropriate. This can derail you from the direction that the evolutionary force is taking you. You must develop discernment and patience, and learn to avoid giving yourself away.

You are capable of developing the capacity to know where you must focus your efforts and your energies. This requires listening intently to what is moving deep within you. Do not allow yourself to be misled by other persuasions.

Overwhelm

The work of food localization is complex and multi-layered, and must occur in the face of great urgency and great resistance, perhaps during periods of considerable chaos and conflict. In such an environment, it is easy to feel overwhelmed and inadequate. Here you must remember that you are not alone in this endeavor, that you are joined not only by those in your initiating group but also by countless others all over the world who are called to this work. And you are mysteriously supported by invisible forces who seek to assist you in aligning ever more closely with the arrow of evolution itself. Eventually, you will be able to feel this, and perhaps even experience moments of grace.

If you find yourself overwhelmed, you must *stop* long enough to recognize the fear and doubt that are holding you back. Recall that you have irrevocably chosen to give all that you are, and do all that you can, in service to a greater purpose. Allow yourself time to reconnect to the deeper current flowing within you. Drink from the well of wisdom and love that lies within your own heart. Summon your courage, and return to work.

Resistance and opposition

While food localization represents healing and regeneration emerging on this beautiful planet, there exist forces that will resist and oppose such efforts.

For instance, the minions of the unholy alliance are legion, and they will mobilize to sabotage, undermine, oppose, thwart, and stop you and your brothers and sisters from being successful. They have a seemingly bottomless war chest, and are prepared for a very long siege. They are prepared to do anything to advance and protect their wealth and power, and will not tolerate anything or anyone who gets in their way. But you are not at war with them. You seek their healing and restoration as well. Remember, it is said that enemies are but friends who have not yet learned to join.

During the time of a dying civilization, the effort to localize the food supply will appear dangerous and threatening to many who are committed to the economic and social status quo. This is an expression of a deep fear that the foundations of society are crumbling, requiring a desperate attempt to shore them up. Be loving and patient with those who are caught up in this drama, and hold yourself apart from it. You know where evolution is going, and you know that it is unstoppable. All obstacles and all resistance are but temporary, and cannot withstand the flow of healing and regeneration.

Capital resources

Building a regional foodshed absolutely requires the redirecting of local capital resources into the emerging food and farming economy. Nearly everyone working on food localization is in the very early stages of learning how to do this.

In Colorado, community economist Michael Shuman estimated that $1.8 billion in capital investments would be needed to get the state to 25 percent local food. Mobilizing such resources seemed an utterly impossible task, until we learned that this represented *less than one-half of one percent* of the money that individuals in Colorado now hold in *non-local* stocks, bonds, and investment funds. $1.8 billion is actually a tiny amount of money.

The challenge is learning how to appropriately redirect such resources. Do not allow yourself to be persuaded that this challenge is an insurmountable obstacle. Innovation is required here, and undaunted persistence.

Remember that real revolutions are not funded by governments, banks, or institutions. They are funded directly by individuals who understand that, with what is at stake, they have no choice but to give what they can.

You and your initiating group must develop the capacity to mobilize capital resources to support the stage of food localization that is attempting to unfold in your regional foodshed. You cannot escape this responsibility.

Sense of urgency

You understand better than anyone in your community how urgent it is to establish a highly localized regional foodshed. You know what is at stake, from the local level to the global. You know that the window of time in which it is even possible to build such lifeboats is rapidly shrinking. But you cannot allow yourself to be driven by desperation. You must remain stable and strong and resilient even when those around you are in a state of collapse.

You need not try to persuade others of the urgency of the situation. You cannot make them feel what you feel, or see what you see. What will be far more effective is the example of your own preparation, the persistent and disciplined body-mind-soul-spirit pursuit of healing and regeneration, and the cultivation of emergence. You yourself, along with your initiating group, must be a thriving center of aliveness.

Making it happen

Partly because you understand the urgency and importance of this work, you will often be tempted to succumb to fear and attempt to exert influence or control, to try to make it happen. You must guard against this tendency, and learn how to restrain this impulse. When so tempted, you must stop and reconnect with your deeper values and commitments. Do not act impulsively, or out of fear, anger, or frustration.

Burnout

In the early days of your journey as a foodshed catalyst, your enthusiasm and naiveté can get you in trouble. You will be tempted to do too much at once, to expect too much of yourself and others. This leads to burnout. Avoid "burning up on the launchpad" of your path, and learn to prepare yourself for a very long journey.

Insurmountable odds

Perhaps the greatest challenge/obstacle you will face is the sometimes inescapable sense that the effort to reclaim the future and take back our food sovereignty faces nothing less than insurmountable odds. There is not enough money, not enough land, not enough labor, not enough support, and not enough time in which to get the job done before the curtain of human civilization falls. This can be paralyzing.

But here you must increasingly realize that you are one with the long arc of the moral universe, and that you carry the seeds of the future that the universe itself has been nurturing for eternity. You have 13.8 billion years of evolution of the universe at your back, and the foundation of countless generations of ancestors to build upon. You are not alone. You are not an individual. Evolution will not fail. Ultimately, you will not fail.

What you can do now

With the perspectives and tools and resources that I have shared with you in this book, you have everything you need to embark on the path of leading the local food revolution in your community. With the plan for preparation that you have developed, you have a solid beginning.

But you may find that you can only get so far on this journey before you realize that it is very difficult to proceed in this work without appropriate support. You may find implementing your plan far more challenging than you had considered.

For those who are called to be leaders in the local food revolution, Lynette Marie and I have created the Local Food Academy, an online bootcamp for emerging foodshed catalysts. Our promise is to empower you to achieve your goals in local food work, whatever they may be, by beginning to take your place as a leader of the local food revolution in your community. The Academy is designed to launch you into a life of being a long-time leader of the local food revolution in the place where you live, and to support you through this journey.

The promise of the Academy is simple: You will be equipped with the perspective, the tools, the processes, and the support that will empower you to become radically more effective in the work of localizing the food supply in your community.

If you are called to this work, this Academy will ignite the local food revolution where it must begin–inside yourself. This Academy is the doorway to shaping your life to become a catalytic leader in the historic local food revolution!

Entrance into phase one of the Academy requires your completing an in-depth application, because participation is limited to a small number of people whom we can work with directly. If you feel you are ready for this level of training and support, and would like to receive details, you can apply here: https://localfoodacademy.kartra.com/page/uVR9.

We are recruiting, training, and cultivating a community of emerging foodshed catalysts who find themselves called to this work. It is our earnest prayer that you will join us.

#

Chapter 12

Conclusion

A s you come to the end of this book, you are approaching the most important moment in your life.

You are now faced with making decisions and commitments that will have far-reaching effects not only in your own life path but will ripple far into the future, ultimately affecting millions upon millions of acres and the lives of many millions of people.

You are among a handful of people in the entire world who grasp the full nature of humanity's predicament and the urgency of radical action. And you are among an even smaller number who understand that building highly localized regional foodsheds is the only strategy that has the potential for forging a pathway to a regenerative future for the human species and for all life on this planet.

These are momentous realizations, and you must be compassionate with yourself as you take them in. Allow time for this, to just be with what you now know—*but not too much time*, for the urgency of the situation compels you to take appropriate action.

You are being called, and the world now awaits your response.

At a time when entropic forces seem to dominate every human activity, when the signs of this civilization's demise are visible everywhere, a revolution is being unleashed. You bear the seed of that revolution in your own heart.

A regenerative revolution demands a decisive response. It is not possible to remain neutral. It is not possible to sit on the sidelines. You know this.

The question now is whether you will answer the call, rising to the greatest occasion in human history, accepting the mission of leading the local food revolution in your community.

Do you have the courage to do this? Are you ready to make an unconditional commitment to this life of service, determined that you will learn whatever you need to learn, develop whatever capacities and resources you may need, cultivate whatever relationships are necessary? Are you ready to join those who are working together to catalyze the local food revolution in every region of the world?

Will you do this? Will you accept this calling?

If so, *when*? You are greatly needed *now*.

If you are ready and willing to take the journey that will ultimately bring you to the place where you can make the most decisive contribution in your community–leading the local food revolution– join the Local Food Academy and declare your commitment along with the rest of us who have already begun this journey together.

We look forward to your joining us in the adventure of a lifetime!

#

Acknowledgments

Lynette Marie Hanthorn, beloved co-creator and co-conspirator.

Dr. Angela Lauria (The Author Incubator), who holds my feet to the fire.

Dedicated farmers and ranchers in Colorado and beyond, foodshed catalysts who are lighting the way forward.

Jean Gebser, Pierre Teilhard de Chardin, Thomas Berry, Brian Swimme, Arthur M. Young, John David Garcia, David Sibbet, John Stewart, Duane Elgin, Daniel Quinn, and Christopher Alexander, who set me on the path of emergence.

Gary Paul Nabhan, Fred Kirschenmann, Jack Kloppenberg, Vandana Shiva, Wendell Berry, Wes Jackson, John Ikerd, Jim Cochran and Larry Yee, Elliot Coleman, Dave Asbury, Allan Savory, Fred Bahnson, Charles Massy, Will Harris, Rich Pecararo, Severine von Tscharner Fleming, Dan Hobbs, Mike Callicrate, Daniela Ibarra-Howell, Mark Winne, Michael Pollan, David Korten, Bill Mollison, David Holmgren, Peter Bane, Mark Guttridge, Mark Shepard, and Joel Salatin, from whom I am learning to see.

Helena Norberg-Hodge, Richard Douthwaite, Richard Heinberg, Julian Darley, Asher Miller, Patrick Holden, Patrick Murphy, Philip Ackerman-Leist, Douglas Gayeton, and Rob Hopkins, pioneers of localization.

Woody Tasch, Michael Shuman, Charles Eisenstein, Judy Wicks, Michelle Long, Vicki Robin, Vicki Pozzebon, and Tom Abood, pioneers of restorative local economics.

Rachel Carson, William Catton, Jared Diamond, Lester Brown, Michael T. Klare, Margaret Wheatley, Timothy Bennett, Elizabeth Kolbert, David Orr, Annalee Newitz, Wen Stephenson, Joseph Tainter, Dr. Albert Bartlett, Bill McKibben, Al Gore, Naomi Klein, Wenonah Hauter, Gus Speth, Peter Senge, Tom Atlee, Sharon Astyk, Dale Allen Pfeiffer, Alastair McIntosh, Clive Hamilton, Lisa Randall, Amitav Ghosh, Dmitry Orlov, Guy McPherson, Micah White, Otto Scharmer, Derrick Jensen, Paul Kingsnorth, James Howard Kunstler, David Grinspoon, Edward O. Wilson, Chris Hedges, John Michael Greer, Joanna Macy, Dan Armstrong, Michael Dowd, Michael C. Ruppert, Carolyn Baker, and Andrew Harvey, harbingers of the Anthropocene.

David Battisti, Peter Wadhams, Kevin Anderson, Paul Beckwith, and Irakli Loladze, courageous scientists dedicated to communicating the difficult truth about climate change.

Dave Gardner, Chef Daniel Asher, Dr. Nanna Meyer, Tamara Campfield, Becky and Bill Wilson, and Alan Lewis, anchors of the original Local Food Summit.

Aspen Moon Farm, River and Woods Restaurant, Basta, Blackbelly, Dushanbe Teahouse, Cup of Peace, Natural Grocers, Whole Foods, and Lucky's Market, for keeping us well fed locally while working on this book.

Jeff Walker, Navid Moazzez, Jeff Hays, Pedram Shojai, Nick Polizzi, and Joseph Michael, who opened my eyes to the catalytic power of online publishing.

Advisors of the Local Food Academy seed launch, including Tom Bartels, Leila Mireskandari, and Stacey Murphy.

To the Morgan James Publishing team: Special thanks to David Hancock, CEO & Founder for believing in me and my message. To my Author Relations Manager, Bonnie Rauch, thanks for making the process seamless and easy. Many more thanks to everyone else, but especially Jim Howard, Bethany Marshall, and Nickcole Watkins.

A host of allies, supporters, and investors (who prefer anonymity), without whose unflagging support this effort would not be possible.

Sister Miriam Therese MacGillis, whose patient love and support kept me out of serious trouble.

Dariel Blackburn, Ellen Rosenthal, and Roberto Chavarria, pillars of our own initiating group.

The Unseen Ones, who have silently guided my authorship and studenthood.

Marshall Vian Summers, who recruited us.

#

About the Author

Michael Brownlee has been working at the front lines of the local food movement for more than a decade, supporting individuals and groups in building regional food systems. As "the foodshed catalyst," he is the author of *Taking Back Our Food Supply: How to Lead the Local Food Revolution to Reclaim a Healthy Future* (Morgan James Publishing, 2019), and *The Local Food Revolution: How Humanity Will Feed Itself in Uncertain Times* (North Atlantic Books, 2016), a manifesto for localizing the nation's food supply and a strategic guide for those called to this mission. He has been a featured speaker with Judy Wicks, Richard Heinberg, Fred Kirschenmann, Woody Tasch, Michael Shuman, and Helena  Norberg-Hodge. A Colorado native, Michael lives in Boulder ("America's foodiest town," according to Bon Appetit magazine), where he and partner Lynette Marie Hanthorn operate Local Food Catalysts LLC, an online publishing and events company supporting emerging leaders in the unfolding local food revolution. Together they are now developing the Local Food Academy for emerging foodshed catalysts, offering online trainings, workshops, personal coaching, and consulting. Michael and Lynette Marie recently produced the online Local Food Summit, featuring more than 90 pre-recorded presentations and interviews with local food leaders in the U.S. and internationally. They also initiated one of the first Slow Money investment groups in the U.S., Colorado Food Investments, providing low-interest loans to local food and farming enterprises. Previously, they co-founded Local Food Shift Group (formerly known as Transition Colorado, the first officially-recognized Transition Initiative in North America), a nonprofit organization dedicated to building community resilience and self-reliance through localizing the food supply.

#

Further Reading

Rebuilding the Foodshed: How to Create Local, Sustainable, and Secure Food Systems, by Philip Ackerman-Leist (Chelsea Green Publishing, 2013)

The Nature of Order: An Essay on the Art of Building and the Nature of the Universe (volumes 1-4), by Christopher Alexander (Center for Environmental Structure, 2004)

Collapsing Consciously: Transformative Truths for Turbulent Times, by Carolyn Baker (North Atlantic Books, 2013)

Extinction Dialogs: How to Live with Death in Mind, by Carolyn Baker and Guy McPherson (Next Revelation Press, 2014)

Soil and Sacrament: A Spiritual Memoir of Food and Faith, by Fred Bahnson (Simon and Schuster, 2013)

The Unsettling of America: Culture and Agriculture, by Wendell Berry (Sierra Club Books, 1977)

The Local Food Revolution: How Humanity Will Feed Itself in Uncertain Times, by Michael Brownlee (North Atlantic Books, 2016)

Overshoot: The Ecological Basis of Revolutionary Change and *Bottleneck: Humanity's Pending Impasse*, by William R. Catton (University of Illinois Press, 1980 and 1982)

Sacred Economics: Money, Gift, and Society in the Age of Transition, by Charles Eisenstein (North Atlantic Books, 2011)

Dark Age America: Climate Change, Cultural Collapse, and the Hard Future Ahead, by John Michael Greer (New Society Publishers, 2016)

Defiant Earth: The Fate of Humans in the Anthropocene, by Clive Hamilton (Polity, 2017)

Wages of Rebellion, by Chris Hedges (Nation Books, 2015)

Permaculture: Principles and Pathways Beyond Sustainability, by David Holmgren (Holmgren Design Services, 2002)

Crisis and Opportunity: Sustainability in American Agriculture, by John Ikerd (University of Nebraska Press, 2008)

Consulting the Genius of the Place: An Ecological Approach to a New Agriculture, by Wes Jackson (Counterpoint, 2011)

Cultivating an Ecological Conscience: Essays from a Farmer Philosopher, by Fred Kirschenmann (Counterpoint, 2011)

This Changes Everything: Capitalism vs. The Climate, by Naomi Klein (Simon and Schuster, 2015)

Change the Future: A Living Economy for a Living Earth, by David C. Korten (Berrett-Koehler Publishers, 2015)

Coming Back to Life: The Updated Guide to the Work That Reconnects, by Joanna Macy and Molly Young Brown (New Society, 2014)

Hell and High Water: Climate Change, Hope and the Human Condition, by Alastair McIntosh (Birlinn, 2008)

Call of the Reed Warbler: A New Agriculture, A New Earth, by Charles Massy (Chelsea Green Publishing, 2018)

Deep Economy: The Wealth of Communities and the Durable Future, by Bill McKibben (St. Martin's Griffin, 2008)

Dirt: The Erosion of Civilization, by David R. Montgomery (University of California Press, 2013)

Growing Food in a Hotter, Drier Land: Lessons from Desert Farmers on Adapting to Climate Uncertainty, by Gary Paul Nabhan (Chelsea Green Publishing, 2013)

Scatter, Adapt, and Remember: How Humans Will Survive a Mass Extinction, by Annalee Newitz (Anchor, 2014)

Bringing the Food Economy Home: Local Alternatives to Global Agribusiness, by Helena Norberg-Hodge (Kumarian Press, 2002)

Down to the Wire: Confronting Climate Collapse, by David W. Orr (Oxford University Press, 2009)

Ishmael, by Daniel Quinn (Bantam/Turner 1995)

Blessing the Hands That Feed Us: What Eating Closer to Home Can Teach Us about Food, Community, and Our Place on Earth, by Vicki Robin (Viking, 2014)

Holistic Management: A New Framework for Decision Making, by Allan Savory and Jody Butterfield (Island Press, 2016)

The Necessary Revolution: How Individuals and Organizations Are Working Together to Create a Sustainable World, by Peter M. Senge (Doubleday, 2008)

Restoration Agriculture: Real-World Permaculture for Farmers, by Mark Shepard (Acres U.S.A., 2013)

The Great Waves of Change: Navigating the Difficult Times Ahead, by Marshall Vian Summers (Society for the Greater Community Way of Knowledge, 2009)

The Universe Story: From the Primordial Flaring Forth to the Ecozoic Era–A Celebration of the Unfolding of the Cosmos, by Brian Swimme and Thomas Berry (Harper San Francisco, 1992)

The Collapse of Complex Societies, by Joseph A. Tainter (Cambridge University Press, 1988)

SOIL: Notes Towards the Theory and Practice of Nurture Capital, by Woody Tasch (Slow Money Institute, 2017)

The End of Protest: A New Playbook for Revolution, by Micah White (Knopf Canada, 2016)

Food Rebels, Guerrilla Gardeners, and Smart-Cookin' Mamas: Fighting Back in an Age of Industrial Agriculture, by Mark Winne (Beacon Press, 2011)

The Reflexive Universe: Evolution of Consciousness, by Arthur M. Young (Delacorte, 1976)

#

The Local Food Academy

Created by Michael Brownlee and Lynette Marie Hanthorn, the mission of the Local Food Academy is to cultivate and support the local food revolution, a radical expansion and acceleration of the local food movement in impact, effectiveness, and scale.

Our primary strategies include revolutionary education, transformational training and coaching, creative collaboration, innovative consulting, and catalytic media communication.

To ignite this local food revolution, to lift the movement to a whole new level of effectiveness and impact and scale, and to do the heavy lifting everywhere, what is needed is a corps of revolutionary foodshed catalysts who are well-informed, well-trained, self-organized, and very well-equipped to contribute in their communities. Together, these individuals will be the foundation of the local food revolution, a catalytic nucleus of healing and regeneration. Such people will be the front lines of the local food revolution!

These are the people the Academy is designed to support.

Our promise to people who engage in this program is simple: *You will be equipped with the perspective, the tools, the processes, and the support that will empower you to become radically more effective in the work of localizing the food supply in your community.*

For more information and an application, click here: https://localfoodacademy.kartra.com/page/uVR9.

Thank You

Finally, my deep appreciation to you for taking the time and energy to read this book. I hope you experience it as the doorway I intended it to be.

To be in touch directly, just email me: michael@localfoodshift.com.

To stay informed of upcoming webinars, trainings, and events, simply hop over to localfoodshift.com and enter your name and email address.

Next steps

Meanwhile, please share this book with other emerging foodshed catalysts using this link: https://localfoodacademy.kartra.com/page/ LUX7, and post your experience of the book on whatever form of social media works best for you.

And don't forget weighing in on Amazon. Your voice matters!

* * *